Guggenheim Aeronautical Laboratory At the California Institute of Technology

The First Fifty Years

Theodore von Kármán (1881-1963). The Hungarian-born aerodynamicist was associated with GALCIT almost from its inception and created a center of aeronautical engineering research and education at Caltech whose alumni came to occupy professorial chairs and directorships throughout the world. He also served as an important advisor to the U.S. Air Force and to the North Atlantic Treaty Organization, whose Advisory Group for Aeronautical Research and Development (AGARD) was organized on his suggestion. (Courtesy Caltech Archives.)

Guggenheim Aeronautical Laboratory At the California Institute of Technology

The First Fifty Years

F. E. C. Culick
Editor

San Francisco Press, Inc.
Box 6800, San Francisco, CA 94101-6800

Box 6800, San Francisco, CA 94101-6800
Printed in the U.S.A.

ISBN 0-911302-46-8
Library of Congress Catalog Card Number 82-050314

PREFACE

On 17 December 1903 the Wright brothers made the first flight of a powered man-carrying aircraft. Within five years, their development of a practical airplane had placed the United States in the first rank of aeronautics. That position was soon lost as rapid progress was made in Europe. The United States lagged well behind until the 1920s.

The Daniel Guggenheim Fund for the Promotion of Aeronautics exerted a remarkable influence on the growth of aeronautics and was a major force as the United States regained its pre-eminence in aviation. In 1925 Harry F. Guggenheim had sought funds to begin a school of aeronautics at New York University. His father, Daniel Guggenheim, decided himself to subscribe the entire amount, $500 000. Less than one year later, in January 1926, he established the Fund, committing a total of $2.5 million for the general purpose of "advancing the art and science of aeronautics and aviation." He instructed the trustees of the Fund that they should "restrict the work to civil activities" and "avoid duplication of effort with other aeronautical organizations."

Under the leadership of Harry F. Guggenheim as president, the Fund achieved dramatic success before liquidation was completed on 1 February 1930. The influence of the Fund endures through the educational and research activities of seven schools of aeronautics bearing the name Guggenheim. The Guggenheim Aeronautical Laboratory of the California Institute of Technology, or GALCIT, began its life of teaching and research in 1928, twenty-five years after the Wrights' initial powered flights.

To commemorate the first twenty-five years of GALCIT, a small volume was published in 1954. The first chapter contains a short history, probably written by Clark B. Millikan, from which the following paragraphs are taken.

"The story really begins some ten years before the construction of the Guggenheim Laboratory itself. The January, 1917, Catalogue of the Throop College of Technology (which later became the California Institute of Technology) contains the following:

Just as this catalogue goes to press, generous and wise friends of the college have undertaken to provide facilities for research in the science of Aeronautics, with every prospect of the cooperation of the United States Government. A wind tunnel will immediately be built and equipped in the best fashion, and a graduate course will probably be provided for students desiring to specialize in this branch of physics and engineering.

"Between $5,000 and $6,000 was made available for these purposes, and during the subsequent year a small NPL type wind tunnel was constructed having a maximum wind velocity of 40 miles per hour. The Throop Catalogue for the following year contains the first mention of two staff members concerned with matters aeronautical. Mr. A. A. Merrill, one of the very early American pioneers, whose active participation in aviation dates back to the 1890's, appears as Research Assistant; he was given the responsibility for designing, supervising the construction of, and operating the wind tunnel. (He also doubled in brass as Instructor in Accounting.) Dr. Harry Bateman, a brilliant Cambridge-trained mathematician, is listed as Professor of Aeronautical Research and Mathematical Physics. The catalogues of the next few years list a number of aeronautical courses given by these two staff members. However, the number of students must have been very small, and there was no Aeronautics Department, nor were any aeronautical degrees awarded. By the mid-nineteen-twenties the aeronautical activities at the California Institute, into which Throop had been transformed, included only the wind tunnel experiments of Mr. Merrill, assisted by occasional students, and advanced courses in Theoretical Hydrodynamics and Elasticity given by Dr. Bateman from time to time to post-graduate students in physics and mathematics.

"At about this time there were two developments of great importance to the future of aviation. The Daniel Guggenheim Fund for the Promotion of Aeronautics was established, having as one of its principal objectives the stimulation of advanced teaching and research in aeronautics. Second, it was becoming increasingly apparent that Southern California was destined to become one of the country's greatest centers of aviation industry. Dr. Robert A. Millikan, Chairman of the California Institute's Executive Council, realized the potential significance of these two factors to the future of the California Institute, and succeeded in October 1926 in obtaining a grant of $300,000 from the Guggenheim Fund for the construction of a laboratory and the establishment of a graduate school of aeronautics at the Institute. The eminent applied mathematician, scientist, and engineer, Dr. Theodore von Kármán, was brought to this country under the auspices of the Guggenheim Fund and visited many educational and research institutions with aeronautical interests. In particular, he spent the fall of 1926 at the California Institute advising its staff regarding the educational policies and experimental facilities of the new graduate school and laboratory. During this visit the essential features which characterized their subsequent development were largely worked out under Kármán's leadership. A cooperative arrangement was also made with the Douglas Aircraft Company whereby the airplane design courses would, at least initially, be given by engineers from its staff.

"During the next two years the laboratory was designed and constructed, having as its major research facility a 200 mile per hour wind tunnel with a 10-foot test section; and the instructional course material was developed. In the fall of 1928 the laboratory was completed, and the GALCIT began its active career in aeronautical instruction and research. The academic staff consisted of Professors Harry Bateman and Theodore von Kármán (the latter as Research Associate, dividing his time between Aachen and the Institute), Assistant Professors Arthur L. Klein, Clark B. Millikan, and Arthur E. Raymond, and Instructor Albert A. Merrill. Raymond, then an engineer and later Vice-President of the Douglas Aircraft Company, served in a part-time capacity and was responsible for the aircraft design courses. Two years later Dr. von Kármán became Director of the GALCIT on a full-time basis and assumed the continuing leadership of its educational and scientific program."

On 15 December 1978 more than four hundred alumni and friends of GALCIT gathered at Caltech to celebrate the 75th anniversary of the Wright brothers' first powered flights and the 50th anniversary of GALCIT. A symposium was held during the day, followed by a banquet in the evening. This book contains the talks given on that occasion.

Most of the talks were accompanied by slides that we have been unable to include here. Some illustrations representative of the texts have been reproduced. Wherever it seemed that no meaning would be lost, references to pictures have been edited out of the written versions.

Transcription from tapes to paper was done by Jacquelyn Beard, Dorothy Eckerman, Karen Valente, and Marcia Hudson. I am especially indebted to Jackie Beard for her help with the final manuscript and many related tasks.

Dr. Judith Goodstein provided several photographs from the Caltech archives and arranged for financial support of editing, through the Caltech Oral History Program. The greater part of the editing was done by Catherine R. Bowers.

For their generous donations defraying the costs of publication, we are most grateful to Dr. F. L. Fernandez (Caltech Ph.D.'69) president, Arete Associates; Dr. Denny R. S. Ko (Caltech Ph.D.'69), president, Dynamics Technology, Inc.; and Dr J. N. Nielson (Caltech Ph.D.'51), chairman of the board, Nielsen Engineering and Research. Their contributions have made this collection possible.

Finally, I wish to thank Hans W. Liepmann for his encouragement and support during the long period of preparation.

F. E. C. Culick

Pasadena, California, 1983

GALCIT staff under wind tunnel, 1930. L. to r., Walter Tollmien (Göttingen), Reinhold Seiferth, W. H. Bowen, Clark Millikan, Harry Bateman, Theodore von Kármán, A. L. Klein, Frank Wattendorf. (Courtesy Caltech Archives.)

Table of Contents

OPENING REMARKS

Marvin L. Goldberger

Nineteen hundred and twenty eight must have been a wonderful year at Caltech, just eight years after Throop Polytechnic became the California Institute of Technology. In 1928 we started both biology and GALCIT, the Graduate Aeronautical Laboratories of the California Institute of Technology. GALCIT originally stood for the Guggenheim Aeronautical Laboratory, a tribute to the original grant of $300 000 from the Daniel Guggenheim Fund for the Promotion of Aeronautics for construction of the laboratory in 1926. That was money well spent.

If I could promise every prospective donor that his contribution would have such a multiplier effect, my job would be much easier. It is important to recognize these days, when we depend so heavily on the Federal government for research support, that relatively small amounts of money injected at the right time, at the right place, and with the right people can be incredibly effective.

The Laboratory was completed in 1928; and Harry Bateman, the legendary Theodore von Kármán, and young Clark Millikan were among the staff members. The Caltech program in aeronautics has always been characterized by its ability to project the needs of future technology—for example, to prepare a generation ready to undertake aircraft industry and national defense needs. Graduates of GALCIT have filled professorships in the leading institutions of the world and positions at the highest levels of industry and in government laboratories. After half a century there is no sign of any loss of vigor.

Dr. Goldberger is president of the California Institute of Technology.

THE GUGGENHEIM FUND AND THE ORIGIN OF GALCIT

Hans W. Liepmann

The origins of GALCIT go back to the 1920s, to what can only be described as a somewhat disastrous state of aeronautics in this country, when the larger part of aeronautics research and development was still in Europe. This fact was recognized by a few people, including Harry Guggenheim, who got together with Alexander Klemin at New York University to raise some money for aeronautics education. Harry Guggenheim approached his father with the idea that led to the establishment of the Daniel Guggenheim Fund for the Promotion of Aeronautics. They mentioned their plans to the President of the United States, Calvin Coolidge, who made a typical remark. When they told him about the future of aeronautics and aviation for rapid transport, he said it didn't help much to get there faster if you had nothing to say. His successor, Herbert Hoover, was far more interested in getting there faster and much more helpful. The Guggenheim Fund was then set up to "promote aeronautical education and the extension of aeronautical sciences, and to assist in the development of commercial aircraft in business, industry, and other economic and social activities of the nation."

That was the original intention and Guggenheim proceeded to endow seven universities: New York, MIT, Caltech, Stanford, Washington, Michigan, and Georgia Tech. Each received an endowment ranging from \$100 000 to \$500 000; New York University was the major recipient and Caltech somewhere in the middle. If you count all the contributions to education from the Guggenheim Fund it amounted to something of the order of \$2 million—less than 10% of the cost of a single 747 airliner today. Even after we multiply it by 10 to allow for inflation, it is an unbelievably small amount if we consider that it was the beginning of U.S. supremacy in aeronautics research, particularly for commercial and military aviation.

Dr. Liepmann is Charles Lee Powell Professor of Fluid Mechanics and Thermodynamics, and director of GALCIT.

Board of Guggenheim Foundation for the Promotion of Aeronautics, 1928: Standing, l. to r., J. W. Miller (secretary), F. Trubee Davison, Elihu Root Jr., H. I. Cone, Charles Lindbergh, Harry Guggenheim, R. A. Millikan; seated, John Ryan, Daniel Guggenheim, Orville Wright, W. F. Durand. (Courtesy Caltech Archives.)

Guggenheim stated clearly that his endowment was not intended to be permanent, but that he thought aeronautics, then in its infancy, would be raised to a state where support would be guaranteed from both private and government sources. And I can only echo Dr. Goldberger's remark that it would have been absolutely out of the question to raise that money for that purpose at that time from the government. Congress would have never permitted such a strange venture. The legislators would have just laughed, because at that time flying was often barnstorming and the accident rate on the few existing airlines was something like one fatality for 1300 miles, a number that has only been reached at times by private automobiles.

Why did Caltech get into the game? We really must admire the foresight of Robert Millikan, who immediately saw the future of aeronautics, and saw that southern California, because of its climate, would be one of the places where the aircraft industry would want to go at a time when flying and flight testing were literally strictly fair-weather undertakings. More important, in his first letter written to Guggenheim in 1925 Millikan stated clearly that the time of the inventor had passed and that the future generation required scientific engineers. He therefore proceeded to find support to bring the best man he could think of to this country to head his school of aeronautics. In his letter to Guggenheim, the three people Millikan mentioned were Ludwig Prandtl, Theodore von Kármán, and Geoffrey I. Taylor. For a physicist who had no direct contact with fluid mechanics or aeronautics at the time, the selection was unbelievably shrewd; moreover, his choice from among the three of the one who certainly should have been chosen was also based on an absolutely correct evaluation.

The Graduate School of Aeronautics at Caltech started with a projected enrollment of ten graduate students and with an unusual emphasis on the scientific foundation of fluid and solid mechanics. This emphasis, which was implied in Millikan's letter and in his choice for the director, certainly paid off. Under Kármán's leadership, the school rapidly became an important source of new ideas and directions in aeronautics. When the war of 1939-1945 broke out GALCIT was practically the only school in the nation equipped to tackle the problems arising from the explosive development of aircraft and missiles to transonic and supersonic speeds and finally into space flight. A list of the approximately 1300 GALCIT alumni (about 10% of all Caltech alumni) reads like a Who's Who of aeronautics; and they have spilled over into other engineering endeavors in which advanced fluid and solid mechanics are important. The offspring of GALCIT, the Jet Propulsion Laboratory, has become a leader in interplanetary exploration. The Guggenheims have good reason to be proud of their child.

THE ORIGINS OF 17 DECEMBER 1903

F. E. C. Culick

I believe I enjoy a unique advantage among the speakers today. As far as I know, there are no eyewitnesses to contradict what I have to say. For most of us our love affairs with aeronautics probably began with aviation in the 1920s and 1930s, and during the two world wars. Aeronautics before 1915-16 appears as a sort of dark age—we are missing a lot. It really is a very interesting story, leading up to the Wright brothers: Wilbur Wright (1867-1912) and Orville Wright (1871-1949). Our ignorance of the history does them a disservice because we cannot appreciate the genuine significance and depth of their achievements.

My interest in this story goes back a little over a year, after I visited the San Diego Aerospace Museum, which burned down last fall. There was a reproduction of the Wright brothers' 1903 Flyer there, and I began wondering why it looked the way it did. A lot of nonsense has been written about the configuration, with frequent references to the possibility that such an airplane is necessarily unstable, which is not true. The Wright brothers' airplane was unstable as they flew it, but it need not have been, so the question is not *why* is an airplane built that way but why did they choose to build *their* airplane that way? That is what got me interested.

The main peculiarity is that the horizontal tail is in the forward position rather than aft. Otherwise it looks as an airplane should, with the biplane configuration, vertical tail, and two pusher propellers. The pilot in the prone position is a bit unusual. Now we might think, "Well, it's an old airplane; things have changed." But that is not an acceptable answer, as we can see by comparing the Flyer with other aircraft of that time. Among aircraft not designed nor influenced by the Wrights, there were some contemporary machines that obviously look like airplanes. The Antoinette is a well-known example. So the reason the 1903 Flyer looked like that is not just that it was built so long ago.

The fact is that the Wrights were extraordinarily successful with that airplane. To see how successful they were, we need only compare their results with other early flights from 1903 to 1908. On 17 December 1903 the Wrights made four flights. The longest one was 59 seconds. Not until November

F. E. C. Culick is professor of applied physics and jet propulsion at Caltech.

1907, four years later, could anybody fly more than a minute, and then only with an airplane that could not be fully controlled. Nobody else knew how to turn properly.

As early as 1905 the Wrights had developed a practical airplane that could fly nearly 40 minutes, and by the end of 1908 Wilbur was flying in France with some flights lasting more than two hours. Not until they flew publicly and people realized what they had done was anyone else really able to fly. They had learned some things—often called "secrets" in that period up to 1908—which they perfected during the next two years. Nobody else discovered what they had done. They kept their secrets very well and eventually marketed their airplanes as a successful commercial venture.

Only by going back into history can we understand the significance of what the Wright brothers did. The story can be traced to 1752 when John Smeaton (1724-1792), who was concerned with windmills and waterwheels, published a paper showing that the force on a plate in a stream varies with the square of the velocity and with the area of the plate. That is true whether it is a flat plate or a curved surface, and is now a well-known fact. Smeaton included in his paper a table of data which he had not taken; they had been provided by somebody else, a Mr. Rouse. From those data others deduced a coefficient (0.005) proportional to the drag coefficient on a flat plate oriented to a stream.

The value for Smeaton's coefficient turned out to be wrong; it remained uncorrected for 150 years. It was used until 1900 and nobody bothered to check it. In 1792 another Englishman, Samuel Vince, using a whirling-arm apparatus, showed that the force acting on a plate varied essentially linearly with the angle of attack between the wing and the relative wind. So by 1800 the basic laws for the forces acting on a body in a fluid were known.

It was at that time that (Sir) George Cayley (1773-1857) began his work. Cayley is acknowledged to be the inventor of the airplane, at least conceptually. He was the first to recognize that in order to fly properly, one must separate the means of propulsion from the means of generating lift. Until then people naturally had been motivated by watching birds and had designed and sometimes built airplanes by emulating flapping wings. Cayley introduced the idea of fixed wings. He was concerned with longitudinal stability and invented the horizontal tail in the aft position. He was also concerned with lateral stability and he invented the use of dihedral, in which the tips of the wings are raised to help stabilize rolling and sideslip motions. Indeed, he invented what we now consider to be the conventional configuration. The gliders Cayley sketched, starting in 1799, look peculiar in some respects but they obviously resemble airplanes that could be made to fly. He built one glider which, according to reports, carried a boy on one occasion. Beyond that he apparently did not progress.

Practically nothing new was added to aeronautics until the work of the inventor of the rubber-powered model airplane, the Frenchman Alphonse Pénaud (1850-1880). Before he committed suicide at the age of thirty, he had made several substantial contributions to aeronautics. The most significant was that he independently invented the horizontal tail as a means of providing longitudinal stability. He knew very well what he was doing, and the fact that he made the successful model is cited in all the literature on early aeronautics.

Yet I have found only one author who took the trouble to read Pénaud's papers and recognized something even more interesting: Pénaud not only used the horizontal tail, he gave the first discussion of how it functioned to provide stability. It is not a mathematical treatment, but is essentially a description of a small–perturbation analysis of stability and the influence of a horizontal tail. The one author who recognized that contribution was Theodore von Kármán (1881-1963), in his book *Aerodynamics* (1954), based on his Cornell lectures.

The most important direct influence on the Wrights was Otto Lilienthal (1848-1896), the first hang-glider pilot. Lilienthal was a practicing mechanical engineer in Germany who made some basic contributions to aeronautics, not only to hang gliding, for which he is best known. In hang gliding the balance is maintained by a shifting of the weight, but Lilienthal's wings had dihedral, and he used a horizontal tail in the rear for stability. So all Cayley's notions are there and the glider looked like an honest airplane. In addition to his gliding work, Lilienthal made aerodynamic measurements of forces on wings, which the Wright brothers used initially in the design of their aircraft. A follower of Lilienthal was Percy Sinclair Pilcher (1866-1899) in England, who also built some hang gliders with the same general configurations as Lilienthal's monoplanes.

In the United States at that time, the most important man in aeronautics was Octave Chanute (1832-1910), an eminent civil engineer who became a bibliographer of aeronautics. In 1894 he published a book called *Progress in Flying Machines,* in which he collected all the available information on flying machines that had been designed or built up to that time. Chanute also built gliders, following Lilienthal. He concentrated on multiplanes, particularly biplanes, all of which had aft horizontal tails. The biplane configuration was not his idea. He got that from another Englishman, Francis Wenham (1824-1908), who in turn got it from Cayley. Chanute's only significant original contribution was the structure used in making the biplane, the Pratt truss.

The Pratt truss had been invented by the American Thomas Willis Pratt (1812-1875) in 1844, for railroad bridge construction. In the airplane it consists of two planes separated by struts. The vertical struts carry the compressive load. Diagonal wires running from opposing ends of the struts carry the tension loads. This eventually became the configuration used in all externally braced biplanes and in particular was taken over directly by the Wrights. There was never any question about that—they acknowledged Chanute's contribution. Chanute's later biplane gliders embodied all that was known of aeronautics at the time the Wrights became interested.

What first prompted their interest was the news of Lilienthal's death in 1896. He was killed when his glider stalled and he could not recover. His back was broken in the crash and he died the following day. Pilcher also was killed in a crash of his glider. Both gliders had aft horizontal tails, and that is essentially the reason the Wright brothers adopted a forward horizontal tail. It was Wilbur's idea. He was afraid of a low-altitude stall, and felt that with a forward tail he could more easily recover from a stall. That is why the tail is in front in the 1903 airplane.

After learning of Lilienthal's death Wilbur became interested in aeronautics and wrote to the secretary of the Smithsonian Institution. He eventually got all the available literature on flying and flying machines. Of importance here is that the Wrights fit into their contemporary technical scene, which is often missed in the literature about them. That is, they really were knowledgeable and in fact by 1900 they were the most knowledgeable people in the world in aerodynamics and aeronautics. They had started right away educating themselves. At the turn of the century Wilbur wrote to Chanute and initiated correspondence that continued for ten years. From those letters, and the very thorough diaries of the Wrights, we can accurately reconstruct the events leading to the invention of the airplane.

In 1899 Wilbur made his first real contribution. From his observation of buzzards, he discovered that birds do not maintain their lateral balance by shifting their weight. In fact, it turned out that they twist their wings a bit. That is how he got the idea of wing warping. Having been told that, you can go out to watch the birds and you can see them doing it. But if you did not know that, and if you were working at a time when everybody else was convinced that dihedral is sufficient for lateral stability, you would be a remarkable observer to notice that birds differentially twist their wings to maintain their lateral balance.

Wilbur realized lateral balance as wing warping and in 1899 made a very interesting kite. As I said, that was the Wrights' first contribution to aeronautics, the first mechanical flying object with both lateral and longitudinal control. It had a 5ft span and was based on the Pratt truss with vertical compression struts and diagonal tension wires. A horizontal surface was rigidly attached to the center strut. That was the only rigid joint. All other joints were pinned, so that the two wings could slide relative to one another.

Cords attached to the wing tips were crossed and held by the operator on the ground. If the hands were rotated in the same sense, the entire top wing slid forward with respect to the lower wing and tilted the horizontal surface up. That gave some longitudinal or pitch control. By rotation of the hands in opposite senses, the right top wing, for example, could be moved forward of the lower wing, and the left upper wing slid rearward with respect to the lower wing. This action produced twisting or warping of the biplane and a variation of the angle of attack over the span such that the lift force was greater on the right wing than on the left. A rolling motion was created and the kite could be rotated about its longitudinal axis.

Wilbur built that kite and flew it on only one afternoon in August 1899. He convinced himself of the validity of wing warping; and that is really when the airplane began. He tried the kite with both forward and aft tail and discovered that either would work, but he chose the forward tail. He also examined the influence of dihedral.

The next step was to build a kite that was large enough to carry a man and that could be held by restraining ropes. In the beginning a man could lie in this kite and learn how to use the controls. After all, flying is just learning how to use the controls. Wilbur realized that right away, and set about to do it.

They needed a place with steady winds, which is how they came to choose Kitty Hawk on the North Carolina coast. It was a day's train trip from Dayton, Ohio, followed by a boat ride to the Outer Banks. It takes a bit of perseverance to put your research laboratory in a place like that. As a matter of fact, on their first trip there the whole project nearly came to an end because Wilbur hired a leaking boat with an inexperienced sea captain and they were caught in a storm. The world's first aeronautical research laboratory was at Kitty Hawk in 1900. Eventually they moved to an area called Kill Devil Hills, a few miles away.

The 1900 kite-glider, a larger version of the 1899 kite, was first flown as a kite. The brothers spent one month at Kitty Hawk in 1900 and flew the glider untethered for a total of two minutes. During those two minutes, only Wilbur flew, and only for ten seconds. That was their entire flying program in 1900. One important thing they discovered was that the lift they got out of this machine was well below what they had anticipated.

To design their glider, they had used Lilienthal's data, which had been taken with a whirling-arm apparatus. The glider was quite small, with a 17ft span and a total area of about 165 ft^2. The lift-to-drag ratio was about 6, the approximate value for all their aircraft. When the lift turned out to be less then they expected, they began to suspect Lilienthal's results; that was the major reason their later gliders were larger.

The whole program nearly ended in disaster very early in the flights of 1901. The wing had more camber to conform more closely to Lilienthal's airfoils and it turned out that the airplane was very difficult to control longitudinally: the horizontal tail movement was insufficient to control the thing and several stalls led to hard landings. By some very clever flight testing of the wing alone, Wilbur deduced the reason: the center of pressure was moving over the wing with angle of attack in a way that he had not expected. It was known at that time that, for a flat plate perpendicular to the stream, the center of pressure is at the center. As the angle of attack is reduced, the center of pressure moves forward until at zero angle of attack it is at the quarter-chord point; so it moves continuously forward as the angle of attack is changed.

What was not known, and what Wilbur found out in flight testing, was that

Wrights' 1900 kite/glider. (Photograph by Orville Wright.)

Wrights' 1901 kite/glider. (Photograph by Orville Wright.)

the center of pressure on a cambered plate initially moves forward from the center at 90° angle of attack and then eventually reverses and moves backward at some intermediate angle of attack. The Wright brothers were running into that center-of-pressure motion, for which they were unable to compensate with horizontal control surface. It is significant that Wilbur himself was doing the flight testing; he was the pilot, making clever observations. They had discovered the peculiarity of the motion of the center of pressure because they were the first to make measurements of lift and drag on a full-scale flying device.

They corrected their problem by reducing the camber of the wings. King posts were installed on the lower wing. They ran wires from the posts to the leading and trailing edges of the lower wing, and to the centers of the ribs in the upper wing. By tightening those wires they could reduce the camber, and their difficulties. So they achieved a variable-camber airfoil merely by adding some wires and a bit of wood.

The year 1901 was doubtless their most important year because they discovered the main problems they had to solve; they worked out solutions; and they ended up determining the essential configuration of the airplane as they chose to build it. They discovered that the lift continued to be much lower than they had expected on the basis of Lilienthal's work. And they also discovered—Wilbur discovered, while flying—the phenomenon of adverse yaw.

In terms of a contemporary airplane, if one wants to turn the airplane one rolls it. For example, to turn to the right, one deflects the left aileron down, which increases the lift on that wing, and deflects the right aileron up, to reduce the lift on the right wing. As a result, one has a rolling moment that tends to roll the airplane, tilts the left vector, and gives a component of force to pull the aircraft into a turn. Unfortunately, what happens is that because the left wing has more lift it also has a little bit more drag. The right wing has less lift, and so has a little bit less drag. The differential drag force causes a yawing motion that tends to turn or yaw the airplane in the direction opposite to where you want to go as a result of the rolling motion. That's why it is called adverse yaw. The way to correct that is to put on a vertical tail, because if you work it through, you find that the vertical tail causes a yawing moment, more readily if it is movable, in the right direction to counteract that adverse yaw and give you a proper turn.

When Wilbur was test flying his funny-looking glider, he discovered that when he rolled it and went into a turn, the nose sort of skidded off instead of following the turn properly. Some time after their flying in 1901 he and Orville figured out that what they had to do was put on a vertical tail. So it is a combination of the wing warping and the vertical tail which is really the essence of their discovery. They were the first to realize that one had to have control of three moments—roll, pitch, and yaw—about the three axes; and that the only way one can properly turn an airplane is to coordinate the roll and yawing moments. And that is the result of what Wilbur observed while flying in 1901. It is easy enough to experience it in a light airplane. If you are a pilot, you know that if you try turning only with the ailerons, you will find the nose skidding around. To be the first to discover that is very special indeed.

The fact that the Wrights found the lift to be much smaller than their predictions led to their wind-tunnel testing. They spent three months, did about a thousand tests, and gathered enough data to use in designing their airplanes for the next decade. They did very little wind-tunnel testing after that, and few people know that in fact they eventually found out that Lilienthal's data were right. What the Wrights discovered in 1901 was that in the region in which they were operating, which was at an angle of attack of 5° to 10°, there was practically no difference between their data and Lilienthal's.

Then they discovered the reason the lift on their wings was less than predicted: the value for Smeaton's coefficient mentioned earlier. By combining the results of their flight tests and their wind-tunnel tests, they figured that the correct value for that coefficient was around 0.003 instead of 0.005. Without going through the argument, let me just say that because of that difference, they had overestimated the lift they would get by the ratio of 5 to 3, or by 67%. And what they had been forced to do was determine the correct value for the drag coefficient on a flat plate. It just happens that because of the manner the data were treated, they needed that number and that is where they had been led astray.

Also in 1901, Wilbur twice stalled the airplane and recovered. Those incidents convinced him that he had corrected the condition that had killed Lilienthal and Pilcher. After that the Wrights were never inclined to use horizontal tails in the rear until their competition forced them to it around 1910.

The first 1902 glider had a fixed double tail. Again the brothers almost ran into disaster. What happens is that in a turn with a fixed rudder, the vertical tail does kill the adverse yaw. But the Wrights very often ran into conditions when they suddenly wanted to correct the turn or correct the balance. Now when you do that there is a transient period when the vertical tail is still giving the moment in the same direction to correct the original adverse yaw. A sudden flip of the warping or of the ailerons yields adverse yaw in the other direction, which in fact combines with the yawing moment of the vertical tail. That tends to cause the airplane to fall off on the wing and perhaps stall and go in.

The point is that you need a movable tail, which they figured out very quickly. In fact it was Orville's idea. It was Wilbur's idea to connect it to the wing warping and that was incorporated in the 1902 glider: a movable vertical tail, although not independent of the wing warping. That is the configuration of the airplane they used after the early 1902 flights. Again they had spent only a month in North Carolina, but in that time both Orville and Wilbur had learned to fly. They made a total of about a thousand glides.

Once they both knew how to fly, they set about putting a motor on that airplane. They did not spend much time worrying about the propulsion system, because they thought that with all the experience with marine propellers, surely one could design propellers easily and well. They did not really do a good job on the engine because they did not work very hard at it. The engine eventually produced 12 hp and weighed 15 lb/hp—not particularly good even at that time. The engine Langley used, which was the best aircraft engine in 1903, weighed only 4 lb/hp.

After building the engine, they discovered that they could not use existing knowledge to design propellers after all, so they had to do it themselves. That was their last great achievement. Again they went to the literature. All that was available to them was momentum theory, which gave them one important result: for a given power output the efficiency is higher if you move more air. That is why they used two propellers with a large diameter. But more important, they worked out the essentials of blade-element theory, viewing each section of the propeller as a little bit of wing. With some help from their wind-tunnel data, they designed propellers with efficiencies of 70% or more, compared with around 50% or less for all their contemporaries. That is why they were able to fly with a rather poor engine.

During this time, Langley's work was well publicized. Samuel Pierpont Langley (1834-1906) was secretary of the Smithsonian Institution; he had a lot of money at his disposal, a big program, and lots of people helping him, and he had spent at least $100 000 or more over the previous twelve years, which was of course a huge sum of money at the time. Now he was about to make some public demonstrations of his full-scale aircraft, attended by representatives of the government, the U.S. Army, and the press. He had flown small rubber-powered models and a quarter-scale steam-powered model successfully. In October he tried his first full-scale flight and failed. The Wright brothers had arrived in Kitty Hawk in September and of course knew about that failure, knew about his attempts, and unquestionably were caught up in the spirit of a race. Langley's first try had been a failure, but the airplane was not seriously damaged and he was going to try again.

The Wrights were down at Kitty Hawk assembling their 1903 airplane, which they started testing in November. In 1903 they spent three months at Kitty Hawk, unlike their programs of the previous three years, when they had spent only one month each visit. In the middle of November, during engine tests, some propeller attachments failed. Since the Wrights were so isolated, they had to send the parts by mail to Dayton to be repaired. They got them back by the end of November; the mails were evidently more efficient seventy-five years ago than today. After a couple of days' testing, a shaft failed, and this time Orville took both shafts back to Dayton to make new ones. He left Dayton on December 9. On the return trip, he read in a newspaper that on the previous day Langley had had his second failure.

It is significant that the Wrights were seriously attacking the flying problem, doing a proper job of design, and doing the flying themselves. The 69-year-old Langley was designing things, building them, and then hiring somebody else to do the flying for him. He had not realized the central importance of learning how to fly.

Wilbur Wright flying 1902 glider with movable single rudder. (Photograph by Orville Wright.)

First powered flight, 17 December 1903: Orville Wright flying, Wilbur Wright standing by. (Courtesy Smithsonian Institution.)

So in early December, the Wrights knew that Langley was out of it and they were the only ones left. Orville arrived back at Kitty Hawk on December 14. They flipped a coin for the first flight. Wilbur won, took the first try, and failed. It turns out they needed a pretty good headwind to get off the ground. He got off in a weak wind—a weak crosswind in fact—stalled, and pranged the forward horizontal surfaces. Actually, only minor repairs were required. On 17 December 1903 Orville took his turn for the flight that became the subject of the most famous picture in all aeronautics. Only when we understand what preceded that flight can we truly appreciate what a remarkable picture it is.

After 1903 the Wrights spent two years developing the airplane and another two years trying to sell it, which is why they did not fly for two years. They succeeded finally in signing contracts both in France and the United States, and in 1908 Wilbur flew in France publicly and Orville flew in the United States publicly for the first time. They amazed the world because no one had figured out that you needed control of both rolling moments and yawing moments, so no one had been able truly to fly and control an airplane.

Wilbur's last public flight was on the occasion of New York City's Tricentennial late in 1911. For $25 000 he flew up the Hudson River from Governor's Island to Grant's Tomb and back. In May 1912 he died of typhoid fever. Orville lived until 1948, but after Wilbur's death the Wright company cannot be regarded as a leader in aeronautics. The Wrights contributed nothing fundamental to aeronautics after the propeller theory, although they contributed a great deal by example until Wilbur's death. They showed the world how to fly in 1908, and the work they did between 1899 and 1903 is unparalleled in aeronautics.

EARLY NACA

Robert T. Jones

I also have some pictures of old airplanes and wind tunnels that you might think of as a continuation of Professor Culick's history. I cannot personally give you the full picture of the earliest days of NASA's predecessor, the National Advisory Committee on Aeronautics (NACA), since I was actually involved in only a part of it, but perhaps my talk will give you a flavor of what was going on then.

The first airplane in my collection looks as if it might have been designed by Paul MacCready, but it is attributed to Alberto Santos-Dumont (1873-1932). In those days aeronauts customarily went up in balloons, and balloons always had a wicker basket under them, so Santos-Dumont is standing up in a wicker basket. He really should have flown before the Wright brothers because he had a much better engine, a 50hp V-8 engine, but it was a few years later that he got airborne this way.

As to the Wright brothers' engine, I was very interested to see and hear one of these engines run at the gathering at Oshkosh this year. It is remarkable in that it has no spark plugs, no carburetor, no throttle. The intake valves are just spring loaded and are drawn in by suction. The exhaust valves are on the bottom; the can at the top, with a pipe leading up to it, is what serves as a carburetor. They simply dripped gasoline on the hot upper surface of the engine, and somehow or other it found its way to the intake valves. The ignition was low tension—what was known as make-and-break ignition. There were low-tension contacts inside the cylinders, operated by a cam from the outside. Why this engine did not catch fire is a mystery because the exhaust just comes out, there are no exhaust pipes, and gasoline is dripping on top. In a later version they had a somewhat better, much more powerful V-8 engine, and they had shortened the wing span and put the tail in back. In many of their early flights, they had linked the vertical tail to the wing warping, so that they got some favorable yaw; that is a technique that Fred Weick developed later and that is now used on my Ercoupe.

By 1910 the airplane had almost achieved its final form. The Antoinette

---

R. T. Jones is senior staff scientist at the National Aeronautics and Space Administration's Ames Research Center.

monoplane still used wing warping, and a 16-cylinder engine (1910). They used single ignition but I think they believed that if they had many cylinders, it would not make so much difference if a few of them went out; but this idea apparently did not really work out. One picture of Hubert Latham (1883-1912) starting on his trip across the English Channel in an Antoinette monoplane shows his throttle, the quadrant, the wheel on one side that operated the wing warping, and the wheel on the other side that was the elevator—really a rather nonintuitive method of control. He started out before Louis Blériot (1872-1936), who actually made it across the Channel. Unfortunately Latham did not quite make it. Blériot did succeed in crossing the Channel with only a three-cylinder engine, but all three cylinders worked all the way.

Soon attention began to be directed toward commercial applications. A picture of the world's first ten-passenger airplane shows Louis Charles Bréguet (1880-1955), who is famous for developing a range formula, and he is at the control. Nowadays, of course we carry 400 or more people, at approximately the same level of accommodation.

I worked for the Marie Meyer Flying Circus in the late 1920s, carrying gas for them in exchange for stick time. I can probably claim the world's record for longest time required to solo. My first flight instruction was in 1928 or 1929, and I finally soloed in 1978. The Standard J-1 was something like a Jenny, but somewhat bigger. Those airplanes were not awfully fast, as you can imagine. After a short period with the Flying Circus, I went to the Nicholas Beasley Airplane Co. in 1929. They were just starting production on an all-metal, low-wing three-passenger monoplane. The structure was all aluminum except the steel tube fuselage—aluminum wings, but covered with fabric. It was really a very advanced design, and carried three people with a 90hp engine. There was a lot of enthusiasm at the time; everyone would own an airplane, just as everyone would own an automobile, and much effort went into the development of a plane that could be stamped out and sold in quantities for a low price. Unfortunately, this was 1929, and the whole project had to be dropped.

In describing the early years of NACA, I shall mention mainly the theoretical developments that were made there. NACA was started in 1915 with a budget of the order of $500 and really no activity. Then Langley Field was built up, the Langley Laboratory. To get work in theoretical aerodynamics started the agency engaged Max Munk, who was one of Prandtl's students. Munk came to NACA in the early 1920s, and did remarkable theoretical work; the most noteworthy example was his development of thin-airfoil theory.

There was of course Nikolai Zhukovski's theory of airfoils, and Wilhelm Kutta's; they were more accurate than the thin-airfoil theory, but they were rather complicated. It was not possible to see directly how the center of pressure, the aerodynamic center, the shape, and the pressure distribution were related. Munk assumed that the boundary conditions were to be applied on the x axis; and since the Laplace equation *is* linear that made the whole prob-

Atendees at 1934 manufacturers' conference (NACA's Ninth Annual Aircraft Engineering Research conference) at Langley Laboratory, with P-26 aircraft in 30 × 60ft full-scale tunnel. (Courtesy NASA Archives.)

NACA cowling being tested in 20ft propeller tunnel at Langley Laboratory. (Courtesy NASA Archives.)

lem linear. This insight proved to be very valuable in later years when we had to calculate the properties of supersonic airfoils.

Here is another example of Munk's ability to simplify things. If you are interested in the properties of ellipsoids and you look in Lamb's *Hydrodynamics,* for example, you find many rather long formulas and lots of elliptic integrals and so on. But Munk noticed that the difficult part of the mathematics was involved in just determining a constant, the maximum velocity over the ellipsoid. Once you have that, the velocity distribution over the rest of the ellipsoid is very simple; namely, you just take the maximum velocity and you project it in the direction of the surface at that point. That is known as Munk's rule.

For instance, if you have an elliptic cylinder with a thickness/chord ratio of 20%, the maximum velocity is 1.2 times the stream velocity and you get the velocities over the rest of the ellipsoid just by projection. Then there is also the rule for the virtual volume of the sphere which is 1.5 times the actual volume. There is a 50% increase due to the inertia of the fluid. Well, the maximum velocity on the sphere is 1.5 times the stream velocity. So here we have a simple relation. For the cylinder the maximum velocity is 2, and the additional apparent volume is 1.

Another important development was Munk's slender-body theory, which I later used in developing a slender-wing theory for transonic, supersonic flow. Using his thin-airfoil theory, Munk derived a family of wing sections for tests in the wind tunnel; several were tested. He found that he could devise an airfoil shape in which the center of pressure did not move, even though the airfoil was cambered.

Dr. Henry Reid, who at that time was not a doctor, was engineer in charge of the Langley Laboratory. Nowadays, anybody who is in charge of anything is called a director or an administrator, but in those days we were not even called scientists, because the language had not become so inflated. Reid's title was Engineer in Charge. Nowadays, practically everybody is called a scientist. The annual budget when I went there in 1934 was $690 000, and when you think how much was being done on that, it really is amazing.

Langley Laboratory had a full-scale wind tunnel, a 20ft propeller research tunnel, a variable-density wind tunnel, a 7 × 10ft tunnel, a spin tunnel, a towing tank, and 300 employees. It is hard to believe that the number of experiments and the amount of work that was done amounted to less than $1 million. In 1934 everyone took a 15% cut in pay, and there is a story that a bill was introduced in Congress to abolish NACA, to save money. Fortunately, NACA executive secretary John Victory was on Capitol Hill at the time and alerted all the aircraft manufacturers to wire their Congressmen to vote against this bill, and he succeeded. So NACA survived, but just barely.

We have a picture of the so-called full-scale tunnel, 30 × 60ft, at Langley. At that time, Orville Wright was a member of the National Advisory Committee, and he attended one of the manufacturer's conferences that they held annually at Langley Lab.

My first boss was Fred Weick who developed the NACA cowling, and at this stage was interested in developing a safe, small airplane and improving the controllability and stability, and he let me work on these problems. The job was part of a Federal program to increase employment, the Works Progress Administration (WPA); it was a temporary nine-month appointment, classified as subprofessional. When my nine-month appointment was up they wanted to spread out the appointment, as there were a lot of people who needed work. I was not permitted to take the Civil Service Examination because I was not a college graduate. The people at Langley asked me to put down the various books that I had studied, things that I had read, and the subjects that one needed to know in order to carry out the work I was doing. They would then issue a special Civil Service Examination to find somebody who could take my place. Well, I had worked for several years operating an elevator in the House office building just across from the Library of Congress, and I used to get out the really old books on mathematics, and so instead of studying vector analysis I studied Hamilton's quaternions, which was a sort of grandfather of vector analysis.

At that time, of course, that subject had been forgotten, had been replaced by vector analysis. But I liked to study the really old books, so when they asked me to put down what a person needed to know to do this work, naturally I put down that one had to have a knowledge of quaternions, and I also emphasized operational calculus. Now when I got there I knew how to write the equations of motion for the airplane, but I didn't know how to solve them. I had studied several esoteric branches of mathematics but not ordinary linear differential equations.

So I set up the equations of motion, and of course Fred didn't know how to solve them either, but he said we'll just make step-by-step calculations, and have all these sets of three simultaneous equations. We'll take it in 0.1 sec intervals and we'll get the operators to compute the thing step by step with a comptometer. They did that and it was interesting, all these couplings among the various motions. Assuming that the aileron was deflected over to a certain position at time zero, we followed the whole thing along, and then Fred said, why don't we try doubling the aileron deflection and see what happens? So I used the same form and I just put in double the aileron deflection, and the operators went through the whole thing again, step by step. I got the sheets back and looked at the numbers in the sheets, and lo and behold, every single number in there was just exactly twice. After that, when anybody asked, have you a linear equation?, I knew what he meant better than anybody.

At that time, Fred was just developing the modern version of the tricycle landing gear and was trying to sell it to the Douglas Company. The Douglas engineers thought it looked like a good idea, but when it landed on its main wheels and flopped over on the nose, wouldn't the passengers in the back be thrown up against the ceiling? So Fred said to me, why don't you compute that and see if you can convince them that this won't happen. So I did. Ordinarily I think most people would just use $F = ma$, but I had to do it by Heaviside's method. Fred made a small airplane he called the W-1 to test his theories.

He also wanted to simplify the controls. He wanted two-control operation, and I worked on that. The main question was, should one eliminate the ailerons and just fly with the rudder and the elevators? Charlie Zimmerman was in favor of that, for some reason I could not quite understand. The alternative was to eliminate the rudder pedals and fly with the ailerons, possibly linked to the rudder. So I worked on that and found very quickly that it was better to use ailerons and link in a little bit of rudder for favorable yaw. Of course, the Wright brothers had found that out some time earlier.

Another picture of a wind tunnel was published in an NACA Technical Report and shows the development of the NACA cowling in the 20ft propeller research tunnel at Langley. That was Fred Weick's idea.

Eastman Jacobs always created quite a bit of excitement at Langley. He was in charge of the variable-density wind tunnel and developed a series of airfoils. He built his own airplane and flew it. Jake's airplane was powered by a 4-cylinder Henderson motorcycle engine which he evidently built in his backyard. He had no flying instruction and just took off with it after taxiing around a few times to practice; people at the Laboratory got wind that he was in the air so they all came running out to the small field to watch him crash. But he practiced enough in the air so that he was able to bring it down without too much difficulty and the airplane flew around for quite a while.

He is also probably the first person to fly in the eye of a hurricane. At that time he had a Pitcairn mail-wing biplane; he flew over to Norfolk and it was parked there when the 1933 hurricane came up and really deluged everything. We had them regularly there at Hampton. When the wind started coming up, he went out to the field and tied the airplane down hard, with its tail into the wind, and waited for the wind to stop, which showed the hurricane had moved along and the eye of the hurricane was right there. He knew which way the hurricane was traveling, so he cranked the airplane up, shook the water out of the wings and the engine, and took off, following the eye of the hurricane back across Hampton Roads and landing on high ground and saving the airplane.

One of Jacobs's major contributions was the development of the laminar-flow airfoil. It was known that the laminar boundary layer had much less skin friction than the turbulent boundary layer, and it was also known, from experiments in England, that the drag of an airfoil depended primarily on how much laminar flow it had. But I think Jake is the one who conceived the idea of designing the airfoil especially to create laminar flow. He tried to apply Theodorsen's method to calculate the shape and got me to help him try to unravel this, but we weren't able to do that. A little bit later, Harvey Allen and I both worked out a technique for designing a shape to suit a given pressure distribution, and Harvey Allen's method proved to be the most accurate, so he used that to design the early laminar-flow airfoils.

At the beginning he used extreme cases, intended to keep laminar flow back to 70% of the chord. Nowadays, people are not so ambitious and they hope for laminar flow back to 30%, possibly 40%. A comparison of a low-drag airfoil with a wire that has the same drag in pounds shows two things. First, it is an illustration of D'Alembert's paradox: it really works experimentally; and second, it shows how small the shearing stresses really are in comparison with the normal pressures.

Theodorsen finally put the potential flow theory of airfoil shapes on a reasonably firm basis; that is, he devised a method whereby you can, by a direct process, calculate the pressure distribution over any arbitrary shape. Nowadays, I have found that you can get almost anything that is pertinent to an airfoil shape on a little magnetic strip 3/4 in. wide and 3 in. long, with an HP-67 computer.

The wing loading of the Wright brothers' airplane was about 2 lb/ft^2. By the time we got to the Standard and the Jenny it was up to 5 lb, and now with the AN 22 and the 747 we are up to 150 lb/ft^2. It is interesting to try to visualize just what 150 lb/ft^2 means, and the way I like to do it is by thinking of a grand piano. Everyone knows a grand piano is an object that is too heavy to fly. It has a surface loading of about 30 lb/ft^2, so if you stack five grand pianos, one on top of the other, you get the wing loading of the 747, which is obviously too heavy to fly.

But then you have to think that the ambient pressure of the atmosphere at sea level is 2000 lb/ft^2, so it is really not straining the atmosphere but it does take a lot of horsepower. Here is what really makes the 747 fly. You obviously could not do it with piston engines; they just have not enough horsepower. We used to be very proud of being able to get 1000 hp into an airplane, but the 747 has four 50 000hp engines. The gas turbine is really the key; moreover, it still gets 60-70 mpg per passenger.

Comment by Dr. Liepmann. Thank you for this talk. I find myself in disagreement on one point. According to my recollection, you have omitted *one* of Max Munk's discoveries—the discovery of Robert T. Jones.

TRANSPORT DEVELOPMENT: THE DCs

Arthur E. Raymond

The DC series of transport airplanes owed much to Caltech's Guggenheim Laboratory, supplemented by the addition of the Cooperative Wind Tunnel, and to the aeronautics faculty and staff. Let me begin with a summary of how that came about.

In 1928 Robert Millikan came down to see Donald Douglas at the Douglas factory, which was then located on Wilshire Boulevard near 26th Street in Santa Monica, to solicit his aid. The Guggenheim Laboratory was being put into operation, but Dr. Millikan felt that the scientists assigned to it would benefit by the infusion of engineering information from someone who was directly engaged in aircraft design and production. He wanted to know whether Mr. Douglas had anyone in his employ who might be able to come to Caltech at least once a week to give a course in this field to prospective faculty and students. At that time, I was assistant chief engineer at Douglas, and the bee was put on me, a task that I assumed gladly.

I had got the M.S. in aeronautics at MIT under Edward P. Warner in 1921, and was familiar with the course material there. Although I had grown up in Pasadena, I had never gone to Caltech as a student but I did audit a course there, in structures, preparatory to joining Douglas in 1925. I also became attached to Caltech when my son married one of the granddaughters of James A. B. Scherer and George Ellery Hale in 1942.

Beginning in 1928 I left Santa Monica at the end of each week and drove to Pasadena to attempt to transfer to my students some of the knowledge I had picked up in the course of my regular job. I continued to do it until 1934, when my boss J. H. "Dutch" Kindelberger left Douglas to become head of North American Aviation and I succeeded to his position of chief engineer. After a year or two I was teaching two courses, one primary and one advanced. Clark Millikan, A. L. "Maj" Klein, E. E. "Ernie" Sechler, Professor Harry Bateman, and Professor Albert Merrill were all in my first class.

Professor Bateman was a mathematician of renown but to my knowledge never became very deeply involved in aeronautics. Professor Merrill was a cute little man with a beard, rare in those days, who had been operating a

---

A. E. Raymond served as assistant and later as chief engineer on the Douglas commercial series of aircraft from the DC-1 to the DC-8.

small wind tunnel at Caltech and who, with the assistance of Clark, Maj, and Ernie, was building an airplane of sorts which they called the Dill Pickle. Its distinguishing feature was an articulated biplane wing that rotated relative to the fuselage, which was supposed to have a beneficial effect on stability. Eventually it flew but was unable to gain enough altitude to clear a haystack and crashed softly.

In subsequent years, my students included Colonel (later General) Don Putt, who went on to Wright field and the Pentagon; W. Bailey Oswald, "Ozzie," whom I recruited to head up aerodynamics at Douglas when we started working on the DC-1; L. E. "Genie" Ruth, who joined us to assist Ozzie, went on to our El Segundo plant as chief aerodynamicist, was loaned to Howard Hughes to design the tail surfaces for his huge flying boat, joined the RAND Project at Douglas (later the independent RAND Corp.) at its inception, and ultimately became president of Lockheed Missiles and Space Co. at Sunnyvale; and J. R. "Goldie" Goldstein, who became head of the Research Division at Douglas and joined RAND as one of the original group; he remained there until his retirement from the position of senior vice president in 1973.

J. E. Lipp, another of the original RAND coterie, was one of my students, as was Carlos Wood, chief of preliminary design at Douglas in Santa Monica and later in Long Beach. And there were the Clauser twins, Milton and Francis, both of whom worked for Douglas but went on to bigger and better positions elsewhere. There were several military officers, too, assigned to take postgraduate work after time in the service. Maj Klein carried on for me during the week, helping with the assigned work I gave out. I am afraid I became very unpopular with my students, particularly my married students, because my assignments always seemed to take much more time than I had anticipated. Maj would sit in the back of my classes and heckle me by interjecting remarks like "That is not so!" We hired Maj as a roving consultant at Douglas, not only because of his connection with the wind tunnels, but also because he was, and is, an expert mechanical engineer.

Much of the wind-tunnel work on the DC transports was done at GALCIT, joined later by the Cooperative Tunnel. Some of the basic structural design parameters on stiff and monocoque metal wings and fuselages were worked out by Ernie Sechler in the structures lab. The DC-1 had its first flight in July 1933, less than a year after we laid out the first sketch. The first DC-2 flew in May 1934, the year Dutch Kindelberger left; and the DC-3, in December 1935. I continued in charge of engineering for the company until I retired in 1960, by which time the DC-8 was in service. Although my duties extended to all Douglas products, I must admit that my first love was always the DC series.

In 1932 United was flying Boeing 247s, twin-engine metal skin cantilever monoplanes. The 247 cabin was cramped and passengers had to step over the main wing spars. TWA was using Fords with corrugated metal skins, and until the famous football coach Knute Rockne was killed in a Fokker accident in March 1931, Fokkers were made with wooden wings. Both these airplanes were tri-motors and TWA badly needed a more modern fleet. Hence Jack Frye's famous letter, which was sent to several manufacturers and read as follows:

> Transcontinental and Western Air is interested in purchasing ten or more tri-motor transport airplanes. I am attaching our general performance specifications covering this equipment and would appreciate your advising us whether your company is interested in this manufacturing job. If so, approximately how long would it take to turn out the first plane for service tests?

As I have said, we did it in less than a year. After listing such items as passenger capacity 12 minimum, cruising speed 150 mph or more, and cruising range 1080 miles, the specifications ended with "service ceiling, any two engines 10 000 ft, and this plane, fully loaded, must make satisfactory takeoffs under good control at any TWA airport on any combination of two engines." This final requirement was what made three engines appear necessary, though all concerned would have preferred to avoid the nose engine of their Fokkers and Fords with its noise, vibration, smell, and inefficiency, and get clear vision forward for the crew.

The Boeing 247 was in operation but could not meet TWA's specifications; we believed the state of the art had advanced to the point where a more modern twin-engine plane could be designed that would meet them. At least it seemed a good gamble, and since this would be a simpler, cheaper, and all-round better design than the one TWA had proposed, we decided to diverge from the specified tri-motor and submit a twin. When we had sketched out a design and made a few calculations as to weight, performance, cost, and delivery schedule, general manager Harry Wetzel and I were delegated to go to TWA headquarters in New York—I, to cover performance and design, and he, to deal with the business aspects with the top brass.

We decided to go by train (since we really wanted to get there), and we used the time in refining our figures. One of my students at Caltech had been "Ozzie" Oswald. In addition to taking my regular course in aircraft design, Ozzie had gone on to my advanced course, and I had given him as a thesis subject the problem of developing an improved analytical method for calculating performance by slide rule. In my discussions in New York, I used Ozzie's method, which became NACA Report No. 408.

I dealt primarily with Lindbergh, then technical advisor to TWA, and our discussion centered on the crucial question of flight with one engine dead. By this time, the simple "takeoff and fly" statement in the Frye letter had been more explicitly defined as a demonstrated ability to take off with full load at Winslow, Arizona, continue safe flight under good control after failure of either engine during takeoff, and land at Albuquerque, New Mexico.

After several days in New York, I was asked to proceed to TWA's main base in Kansas City, there to confer with Jack Frye and his operations and maintenance people regarding the writing of detailed specifications. I am not sure at what point we were informed we would be given a contract. We had two competitors, General Aviation and Sikorsky, both of whom submitted tri-motors.

There is some evidence that, to cover its bet, TWA gave a second contract to General Aviation, which was later cancelled when rising confidence in the DC-1 overcame doubts.

As a matter of policy, and to learn more about the current operation of the airline with which we were dealing, I decided to be brave and fly to Kansas City rather than to take the train. So I boarded a Ford tri-motor at Newark around 9 a.m. We bumped along all day at low altitude and finally crossed the Mississippi at dusk. I remember thinking how miraculous it was to have gone from the Atlantic Ocean to the Mississippi River in one daylight period. Landing on the wet field at Kansas City after dark, I got a blast of muddy water in my face from the S-shaped ventilators that brought fresh air directly from outside. After that tiring, unpleasant flight I could see why TWA needed better transportation.

The specifications I brought back to Santa Monica, a copy of which I have with my pencil notations in the margin, was 15 legal-size pages. By then I was sure we were faced with a challenge, so about the first action I took was to hire Ozzie, temporarily, I told him, for two weeks; but of course he, like Maj, stayed until retirement age. I asked him to concentrate on aerodynamics, and particularly to confirm or improve the calculations I had made regarding single-engine flight. To tell the truth, I was a bit uneasy about that myself.

We had good men in that original DC-1 group. Among them were Ed Burton, who became for many years chief engineer of our Santa Monica plant, finally succeeding me when I retired in 1960; Lee Atwood, who did stress analysis and other design layouts; and Ivar Shogran, who handled the power plant work, to name a few. I wish I had a complete list, but the only one that I

DC-1 (1933) at Glendale's Grand Central Air Terminal. (Courtesy Douglas Aircraft Co.)

have found is dated 1935, by which time there were far too many to mention. Lee Atwood, as you know, eventually succeeded Dutch Kindelberger as president of North American.

Jack Northrop, who worked for Douglas shortly after I joined, left to form his own company and to develop monocoque construction, first in wood and then in metal. He had come back to form a Northrop subsidiary to Douglas, later to become the El Segundo plant; the famous test pilot Eddie Allen was working for him. Eddie came over and joined Ozzie in establishing a system for calculating flight plans for greatest efficiency. This system clearly showed the advantage of flying higher than was then customary. Since a higher flight was usually a smoother one, this soon became standard practice. Emphasis on supercharged engines was also a natural result. Ozzie and Eddie made a good team.

The DC-1 had the advantage of being able to combine a number of recent advances in the art: a Northrop type with monocoque construction as applied to duralumin fuselages and cantilever wings, the NACA enclosed type of engine cowl, efficient wing flaps, and adjustable metal propellers. It turned out, not unexpectedly, to be overweight, but during its construction a two-position controllable-pitch propeller became available and there was an increase in approved power owing to the introduction of higher octane fuels, and these advances did much to compensate.

The swept-back wing, by the way, was not a part of the original design, nor did it come out of the wind-tunnel tests. The fact is that as the weight grew, the balance moved aft. The location of the center wing on the body had already been fixed and it was easier to sweep the outer panels than to move the entire wing. That at least gave the plane an easily recognizable silhouette.

Carl Colver, then sales vice president, and Fred Herman, project engineer, were pilot and co-pilot on the first flight. Those of us who watched were treated to a thrill. Takeoff was toward the ocean. In that direction the land beyond Cloverfield drops off rather sharply. The airplane took off nicely, but after it gained a bit of altitude it began to sink as it went into a left turn, to the point where it dropped out of sight. Our hearts nearly stopped, but then as the DC-1 continued to turn we saw it again, gaining some altitude, only to sink and again rise; and this time it remained in sight. The third time this happened we relaxed a bit because Carl was obviously able to get a little higher each time. He made a normal landing on the field after a 360° turn.

In fact, he made a second flight with our flight engineer Frank Carbone as co-pilot and did considerably better. But after landing Carl said to Ivar Shogran, the power-plant engineer, "This time, I'd like you to come along and see for yourself what's happening rather than telling you about it." After this flight, there was general agreement that it was probably carburetor float trouble. Upon inspection, it was found that the carburetor attach bolts were positioned so that it could be turned 180°. That was done and the problem was solved. Actually, the carburetor had not been improperly installed; it was a new type, and although thoroughly checked on the test stand at Wright, had never been subjected to the varying accelerations and angles of flight.

While I am on the subject of engines, it might amuse you to know about the competition that developed between Wright and Pratt & Whitney, later in the DC-1 test program, regarding engine choice for DC-2s. We divided part of our shop into two areas, one for each company. Shogran tells me that with experience, it became possible to change the airplane from one set of engines to the other in half an hour. Both makes of engine developed their own sets of troubles, so that they were removed and replaced alternately several times for modification. The most serious trouble with the Wright engine was overheating—and believe it or not, Wright was able to design, build, and deliver improved cylinders ready for flight, all within 10 days. By this time, the respective engine crews had been fitted with distinctive T-shirts like football teams, and as soon as one team came off the field, so to speak, the other was ready to rush into action.

TWA chose Wright engines for their DC-2s and so did practically everyone else. About two-thirds of the commercial DC-3s had Wrights; the others, an advanced version of the Pratt & Whitney (P&W). All the military versions of the DC-3s had P&Ws, so P&W came out ahead in the long run, as they had a habit of doing.

We had several problems with the DC-1. Eddie Allen, experienced as he was, made a wheels-up landing, bending the propeller tips, so we improved the warning system, Actually, he was not altogether to blame because in those first flights the respective tasks of the three crew members were not well defined, and two members of the crew each thought that the other was supposed to let the wheels down. Ivar tells me these propellers were the only three-blade controls in existence at that time, so instead of installing new blades we returned the bent ones to the factory in Hartford to be straightened out, and the plane was flown with fixed-pitch propellers until they came back.

The landing gear had a disconcerting tendency to collapse of its own accord, and we had to develop an improved latching arrangement to keep it down when it was supposed to be down. At first it had to be pumped up and down by hand; later we installed a hydraulic pump. Originally the flaps were also manually operated, with a long lead screw with which it took many turns and a long time to lower them, so we powered them also. As was usual in those days, some modifications in the tail surface were necessary to improve stability and control and to achieve proper stick forces, but these changes were minor. The stalling speed came out a bit high and we tried an auxiliary wing between the nacelles and the fuselage to see if that would help. It did lower the stalling speed but made the stall so abrupt that we decided against it. Other than that, practically all the performance specifications were met or bettered.

The single-engine test was successfully made, starting at Winslow (4900 ft), passing over Gallup (6500 ft) and the highest point on the TWA route (8000 ft), and landing at Albuquerque (5300 ft), a distance of 240 miles. Eddie Allen was the pilot, Tommy Thomason was the co-pilot representing TWA, and Frank Carbone was the flight engineer. As soon as the plane was free of the ground, the switch on one engine was cut and the landing gear retracted. The plane did sink a bit before starting to climb, but the propeller tip stayed clear and from then on all was well. The DC-1 was ballasted to full gross weight, 17 500 lb.

The DC-1 cost about $300 000, but its price to TWA was $125 000, as was typical in those days. They also exercised their option at below cost, but that did not bother us very much because by that time we knew we had an airplane that could be sold to others in quantity. However, we did not then realize its full potential. After 1933 the DC-1 had a long and checkered history. Suffice it to say that after fully earning its laurels and passing through many hands, it finally cracked up in a noninjury accident in Malaga in Spain in 1940.

The production DC-2s that followed the DC-1 differed mainly in that they were stretched to increase the normal passenger capacity from 12 to 14. There were improvements throughout as dictated by experience and availability of advanced engines and equipment. The first DC-2 flew in May 1934, 10 months after the DC-1, but it really caught the world's attention when one flown by KLM came in second in the London-to-Melbourne race for the MacRobertson trophy, in October 1934, operating as an airline airplane with paying passengers. The winner was a souped-up fighter and it did not win by much. We sold around 200 DC-2s, but before they were all delivered the DC-3 was in the air.

The DC-3 flew in December 1935 and was in service six months later. The DC-2 had begun the process of putting commercial air transportation on the map as a profitable and reliable business. Early in its service career, I can remember coming over La Guardia Field in New York on a delivery flight one night, when our pilot had to circle several times and gun the engines to get the field lights turned on. That situation did not last very long. In the two years between the London-Melbourne race and the advent of the DC-3, air travel came of age.

With the DC-3 it really began to flourish. This airplane started as the Douglas Sleeper Transporter or DST, born of a requirement by American Airlines for a better air Pullman than its Curtis Condor biplane. This requirement called for rounding out the fuselage to accommodate berths, and the increased width made it adaptable to normal three-abreast seating or crowded four abreast. Normal growth in engine power and gross weight, together with detailed improvements throughout the structure and installations, resulted in other changes; but the wider fuselage and increased payload were chiefly responsible for the great acceptance and demand that greeted this model. It had no competitor for years and established Douglas as the preeminent builder of transport airplanes for more than 20 years.

Moreover, what would have been merely a truly remarkable civilian market turned into a fantastic overall demand when the military requirements of

World War II were added. All told, we built close to 11 000 of this model, though the exact number seems elusive to pin down (as is the number still in service). When we started production, we had to figure out the quantity that we expected to put through the tooling, and we decided that we would be really brave and instead of designing a tooling for 25 airplanes, design for 50.

American Airlines was the initial DC-3 purchaser, and because they had an extremely capable chief engineer, Bill Littlewood, I gave them almost a free hand in establishing the dimensions of the cabin and deciding what went into it and in the cockpit layout. American and Douglas worked together with little friction. Thus, the DC-3 was the product of teamwork between the builder and a major user, one of the reasons it was so successful. Another reason was its extremely simple structural design, with few concentrations of load, which made its life virtually endless because replacement of individual elements was so easy. Finally, it had the advantage of following the DC-1 and 2, so that by the time it came along we had worked out most of the bugs.

Several years ago, I had the pleasure of flying in a DC-3 that had logged more than 80 000 hr. Engines and tires had been replaced several times but the plane looked like new.

Of course, since it was unpressurized, passengers would have to be supplied with oxygen when it cruised above 12 000-14 000 ft. We anticipated building a pressurized DC-4 with four engines, but in the meantime we experimented with various types of oxygen masks, none of them very satisfactory. So we conceived of the idea of filling the entire cabin with oxygen-enriched air and tried it out on the DC-2.

We filled the plane with a miscellaneous group of passengers made up of members of the board of directors and their wives, children, and friends. We wanted a cross section of humanity, and we took along the company doctor to take their pulse, blood pressure, and respiration. Frank Collbohm established

U.S. Army Air Corps version of DC-3 during World War II. (Courtesy Douglas Aircraft Co.)

DC-7C (Seven Seas), last reciprocating propeller-driven four-engine Douglas transport. (Courtesy Douglas Aircraft Co.)

DC-9 (1980) two-engine jet transport. (Courtesy Douglas Aircraft Co.)

a station in the front of the cabin with a tank of liquid oxygen and means for dispensing it. The risk of explosion or fire was not given much thought. We flew over Boulder Dam at 25 000 ft and experimented with lighting cigarettes to see how large the flame was. We had to scratch the frost off the windows to look through them. Luckily we emerged unscathed, but I have often thought when looking back on this episode, how naive can one be? We did show enough sense not to adopt this arrangement for production.

I look back on those days with much nostalgia. In a way, they represented adolescence at its best. We had a sense of being mutually engaged in something worth while. It was fun and it was creative, and the future looked bright, as indeed it was. I imagine those who worked on the Gossamer Condor, which is the subject of the next talk, must have felt much the same way. We were doing things for the first time and there was a very special intimacy between us and the work of our hands. As time went on, we became involved in much larger, more complicated, and more costly airplanes, and as individuals we inevitably became more specialized and less involved in the product as a whole. I imagine that the comparatively small group who worked on the DC-1 knew every bolt, nut, and rivet that went into it, so to speak. We also felt very close to each other and were in daily constant contact. It was a happy family and a happy ship.

Let me close by giving you a few comparative figures. First, how well did the DC-1 meet its promised performance and weights? The gross weight called for in the Kansas City specification was 14 600 lb. We did not do very well on that: the actual weight was 17 500 lb. We did better on speed, thanks in part to the higher power of the engines, 710 instead of 600 hp. Nowadays, we would have had to change the required specifications to take care of the increased horsepower, but then it made no difference. The attained high speed was 210 mph instead of 183, and the cruising speed 187 instead of 150. Cruising range was 1200 miles, up 200 miles. We were a bit over on landing speed and a bit under on single-engine ceiling, but field-length requirements were satisfactory, and we demonstrated the ability to fly on one engine over Gallup with plenty of altitude to spare.

Second, in comparing the DC-1 with the early DC-3s, engine power increased from 710 to 900 hp. Gross weight was 24 000 lb as opposed to 17 500 lb. Speeds were about the same. Both the cruising range and one-engine ceilings were improved, from 1200 to 1400 miles, and from 9000 to 10 000 ft. Passenger capacity went from 12 to normal 21.

Third, here are some figures on the latest McDonnell design for a twin-engine transport, the DC-9 super 80. Gross weight, 140 000 lb; maximum payload, 40 200 lb; maximum speed, 600 mph plus; cruising speed, 520 mph; passenger capacity, 155 typical, 172 maximum; range with maximum payload 1500 miles (2050 miles with 137 passengers), operating ceiling 35 000 to 37 000 ft (variable with fuel used), ceiling on one engine 13 000 ft. So we have come up a long way on two engines in 43 years, but the DC-3s are still operating.

THE GOSSAMER CONDOR

Paul B. MacCready

I shall tell you the general story of the Gossamer Condor, and then Dr. Lissaman will follow with more details on the technical points and a bit more on the overall design philosophy, to put that in perspective. Development of the Gossamer Condor aircraft was a fantastic happening for all of us who were involved in it. In retrospect, it seems as though everything in my background constituted an element of preparation—model airplanes, sailplane flying, the aeronautics department at Caltech, daydreaming about flight vehicles when I should have been listening to Homer Joe Stewart or Clark Millikan during courses, acquiring Peter Lissaman and Jim Burke as friends and associates, becoming involved deeply with my sons in the growth of hang gliding, and being in southern California, which is really the place to be when it comes to aviation.

The Kremer prize seemed made for this particular combination of elements in one's background, though for seventeen years I certainly did not suspect that was the case. In one way, the truly creative part was Mr. Henry Kremer, the British industrialist, putting up the £50 000 prize, aviation's largest single award. He rescued the subject of human-powered flight, which really has no practical applications nor economic rewards, and gave it a feasible goal.

The basic idea of the Gossamer Condor is as simple as can be. The power consumed by an airplane, a glider, is its weight times its sinking speed; the sinking speed, the gliding speed, is proportional to the square root of the wing loading. You put those elements together and you find that power is proportional to the weight to the three-halves power for an airplane of given shape and inversely proportional to the span.

Now in the case of the Gossamer Condor, or any large hang glider, the weight does not depend strongly on the size. The weight is primarily the pilot; the rest of the vehicle is fairly light, whether it is big or small. That means that the weight is fixed and we are back to power being inversely proportional to scale or to span; so if you have to get down to a certain power we just keep making the plane bigger and bigger, without the weight going up, and we can always get down to any arbitrarily low power we want. It may end up as an im-

Dr. MacCready is president of AeroVironment Inc., and the designer (with Peter Lissaman) of the first successful man-powered airplane, the Gossamer Condor.

practical sort of airplane, but then again it may not.

In this particular case, I was able to do the calculation quickly, extrapolating from what I knew about hang glider performance. Where a 30ft hang glider, a really efficient one, flies at about 1.2 hp, when we triple all its dimensions it can fly at about 0.4 hp, and that is in the range that a man can produce. So that was the basic idea, estimated in one's head. Then one started doing some calculations, and it turned out that probably an 80ft span would do the job. To give some margin, we made the span 96 ft instead so that we wouldn't have to have everything just perfect. We could get by with some parts not working the right way as long as we had an extra margin of size.

I remember one of my biggest worries was, would all these pieces of wire that we needed—the piano wire that is holding everything—would they have too much weight or drag? In doing the calculations I was delighted to find that for the range of Reynolds number at which we would fly, the drag coefficient was still only around 1 or 1.2; it turned out that these all-important wires only represented 1-2% of the weight of the vehicle and perhaps 5% of the parasite drag, so they let the thing be feasible.

Simple theory, of the sort I have described, predicted how big the wires should be. The rest of the project consisted in deciding how to let that wing fly stably, using the minimum weight and the minimum of everything else. One needed the stabilizer because the wing by itself had not enough pitch stability. One could put the stabilizer either in front or in the rear. For the particular kind of wing construction we were using, we needed some bowsprit sticking out in front anyway, so without adding any additional weight, we could stick the stabilizer there. That was really the first reason it was put in front, but there was a bonus: if the stabilizer is in front the propeller can be in the rear, which is a particularly efficient place for it. So the design just went along until the glider looked right.

You can build a crude hang glider in one weekend because it is so simple—pieces of tubing bolted together and cables. Cables are really the structural key here, tubes in compression and cabling to hold it all together; and cables can take a pretty good load. I was familiar with hang gliders, had done a lot of thinking about them, comparing them to sailplanes and also to soaring birds when I was on vacation ridding my mind of business worries. I was thinking about the wing loading of birds and the amount of power they use versus the wing loading and power of hang gliders.

Incidentally, you can study the flight of soaring birds flying in circles even while you are driving along in a car. You measure the time to turn and estimate the bank angle, and that gives you their speed and the radius of turn; and if you look up some data on the wing loading of the birds, you can even estimate the lift coefficient at which they are flying in the circling flight. When I did that it came out to 0.9, which is a very reasonable value, not some superhigh value as people have suspected. So, these kinds of things were milling around in my mind and finally came out with the design that I just mentioned.

After the first design, coming back from vacation, I built a couple of little models, and then we built an 88ft model. We flew this model in the rain in the middle of the night, but it was enough to show that the structural concepts were reasonable and the pitch stability was adequate. The testing involved running along holding it like a big model airplane. We then took it out to Mojave Airport and rebuilt it several times. The first significant flight was the day after Christmas 1976, when the earliest version of the Gossamer Condor actually flew. It was a flight of about 40 sec.

We had had many short flights before that, with the vehicle in slightly different configurations, and found that the apparent mass was dominating some of the flight activities. The apparent mass for the vehicle was three times the mass of the whole vehicle and pilot, something with which one is never confronted in ordinary airplanes. If one tries to accelerate or turn, it takes a huge amount of power, but at just the right speed it would go along smoothly.

There were many small accidents. For example, in a hard landing the vertical post might break. Then we might strap on a broom handle with some duct tape and be flying again a few minutes later. We would fix up the tubing professionally that night for more flights the next day. The same sort of accident on a complicated aircraft, such as those which had been constructed in England, would require several thousand man-hours to fix up, and sometimes the fixing up would not take place and the whole program would be aborted.

Finally the plane began flying better and better, and we got up to 2.5 min with it at Mojave. But it was obvious that plane was never going to make it around the Kremer course, which is a figure eight-course around pylons half a mile apart, involving some climbs and turns in both directions, and the flight long enough so that it defines sustained controlled flight. So we did a lot of thinking about the structure and aerodynamics and how to make this thing turn, because this one just would not turn. If one were to put a big enough force, a spoiler on the wing tip, to get a little turn out of it, the drag would go way up and the vehicle would just sink to the ground.

The apparent-mass effects really were bothering us. We found it beneficial to do some crude tests with a wood model in water. One can push the model around and have large apparent mass effects that are not there with ordinary little models flying in air—nothing quantitative, yet one does get a genuine feel for some of the factors. Other clues came from computer studies on stability and control, by Peter Lissaman and Henry Jex. Thus, a good bit of computer work and a lot of hard thinking went into trying to assemble all the clues we had from the little model tests and our flight tests.

It became obvious to us that one cannot make the airplane roll just by pushing up or pushing down on one wing; it is almost impervious to roll that way. Some other tricks must be used. The apparent-mass effect must be brought down a lot, which means that the wing tips must be much smaller. We cut them down from 12 to 5 ft, which gave less area and a higher speed. We rocketed up from 8 to 10 mph, which doubles the power that goes into parasite drag, so we had to make up for that by streamlining on the cockpit for the pilot. We had to get better wing structure by putting the spar near the center of pressure, which meant going to a double-surface airfoil.

The resulting design worked right off the bat. The most thrilling moment of the project to me was pushing it off with my son in as pilot, after dark one night the day we had finished it, and feeling how much easier it was to shove into the air than the previous versions; and then having the pilot report after he landed out in the dark how much easier it was to fly. We knew that this plane was going to do it, but much effort would be required before we would get there. We made a 5min flight with a good professional bicyclist as pilot one day, and thought: even though we are not ready, let's make a try, at least sort of a dry run, at the Kremer Prize flight.

That flight ended in a dramatic, slow-action crash. We had not enough control, for various reasons that we discovered later, and although we could initiate a turn, we could not be sure we could maintain it. The plane was efficient enough to get around the course then, but it could not do the turns. We kept working away on various versions; we replaced the wing, the fuselage, the stabilizer; we worked on various control systems; and eventually, with quite a few man-hours gone into it, we came up with a system of rocking the stabilizer for turn and wing twisting. It all worked and gave magnificent control.

Flight speeds are only 10 mph, which makes flight tests very comfortable. You can run alongside, look at tufts, and talk to the pilot. The pilot and engine was Bryan Allen, an excellent bicyclist and hang-glider pilot. On one flight a few weeks before the 23 August 1977 final Kremer Prize flight, he was in the most dramatic crash. A third of the way into the flight a pitch control line malfunctioned, and the vehicle executed a dive into the ground. It was the fourth time we had a complete crash like that. The plane can be repaired very quickly. Within 24 hours it can be flying again, but we always took the

Greg Miller, pilot, March 1977, Shafter, California. (Courtesy Don Monroe.)

Bryan Allen flying final version of Gossamer Condor, November 1977, Shafter, California. (Courtesy Don Monroe).

occasion to doctor it up a bit. In this case, we had been building a lighter fuselage and were looking for an excuse to put it on. So we put the new fuselage on, saved 6 lb, and recommenced test flights. The previous month every flight had been a little longer than the preceding flight, and we were sure we were close to the final result.

We pushed on takeoff to lessen the distance the wheels had to roll on the ground while the plane had the pilot on board. The little toy wheels we employed would otherwise wear out after only a few takeoffs. The plane becomes airborne in about 2 or 3 sec if you are pushing; otherwise it takes the pilot 4 or 5 sec. The takeoff is obviously no strain as long as the plane is capable of flight. The pilot tried one day and got all the way around except for the last hurdle; he could not get over the last 10ft marker because he did not realize he was so close to it. He did an 8 minute flight.

On the final flight, Bryan went easily over the 10ft marker at the beginning of the flight, and then manuvered all the turns comfortably. The combination of wing twist (via wires) and a tilting stabilizer (tilted aerodynamically by ailerons or tabs at its tips) permits coordinated turns. The stabilizer, when tilted, provides yaw trim; because the stabilizer is lifting, tilting it to one side or the other pulls the nose around.

It was the most carefully observed record flight that has ever been made, in that the observer was within one span of the airplane throughout the whole flight. When everything was going perfectly, and there was no turbulence and no wind, Bryan Allen, who weighs about 140 lb, could fly it on 0.33 hp. During this flight he probably averaged something like 0.42 hp because there was some turbulence. Turbulence really hurts; it adds to the parasite drag and to the induced drag. The turbulence effect is significant whenever the wind is more than 2 mph.

In contrast to various animals, humans walk poorly, swim poorly, and fly not at all. With apparatus, they can make up a little for their general physical ineffectiveness. Their evolutionary development has been in their brains rather than in their muscles. Now we have a human-propelled apparatus that permits flight. During the past decade there have also been advances in the swimming and walking departments. Streamlined bicycles, or more accurately streamlined vehicles based on bicycle technology, have been exceeding 50 mph and getting some publicity. There is a contest for these vehicles at Ontario Motor Speedway in California every May. To help man propel himself in water, there is an underwater device called an Aqueon, designed by Cal Gongwer. The swimmer puts this device between the legs and wiggles them in a dolphin mode; the resulting motion is a lot faster than when swimming in an ordinary fashion. In summary, as one starts to apply some technical concepts to locomotion, one can move very differently from the way animals have been moving, and in some cases faster.

Now we get to the next plane, the Gossamer Albatross. The Kremer Prize had stood for 18 years, and after it was won, Henry Kremer put up a 100 000 prize for flight across the English Channel. I an sure he thought a challenge would last for another 18 years. We had already been designing a plane that we felt could do it, because we wanted a plane to fly once ours had gone back to the Smithsonian Institution. Calculations quickly showed that by improving the construction techniques one can make a much better airplane.

The existence of Kremer's new prize gave us a stimulus to push a design that used carbon filament tubing and styrofoam, and involved much more care in the building. This new plane, the Gossamer Albatross, embodies no new ideas but does feature better structural integrity. It has a nicely shaped wing, almost like a sailplane wing. It weighs only 55 lb instead of the heavy overweight 70lb Gossamer Condor. We were flying it in July and a little bit in August of 1978. It takes much less power than the Gossamer Condor; you just get in and pedal. If flies on the power that would make a bicycle go 19 or 20 mph, which a good bicyclist should be be able to do for 5 hours or so. The cyclist is seated in a different position, which may help a bit in the amount of power he can deliver. We are pretty sure that the Gossamer Albatross can cross the Channel.

However, building the plane and having the right engine for it only represent about 10% of the complexity and cost of the project. We are still working on digging up sponsorship to pay for the huge logistics efforts and so on required in back-up vehicles, boat rental, and everything needed for a safe flight across the English Channel. We expect to do the flight sometime in May or June of next year. (*Note:* The flight was successfully completed at 7:40 a.m. on 12 June 1979. *Ed.*)

It is worth while mentioning some reasons the Gossamer Condor succeeded. It started with the right concept of scale and structure, not burdened by old thinking, and had a performance margin so that we could do things crudely.

The most important part of the project was the team of experienced people dedicated to a clear goal. It was not an individual effort at all. The vehicle was easy to build, modify, and repair, so that in one year we did more development than would have been possible in a decade the normal way. There was a lot of emphasis on flight tests, especially important in pioneering some new realm—in this case, flight under 10 mph. We had complete confidence right from the start that the project would be successful. There were certainly many problems, but we assumed that with every difficulty the simplest and easiest approach would work. In most cases it did, and that left time to focus on the few very troublesome items.

An important element in the whole thing was luck. Some of the things we stumbled on just worked right. With some of the stability aspects, we did something for one reason and solved a problem for a different reason. Stability and control are better than we really deserve. One would not want to build a space shuttle or a DC-10 with the same philosophy, but for this project is was just right.

One wonders what good it is, how significant it is, what it leads to, and whether there are applications? It is viewed by many as a milestone in aviation, which is why it is in the Smithsonian. It was a pioneering development like that of the Wright brothers, but it certainly was not significant in the same way. We do not see any big industry building up from it; it is not going to change the course of aviation or of the world. The man in the street appreciates it because he can identify with it more than he can with a shuttle or an X-15; and people working in large aerospace firms appreciate it because it represents a few individuals going out and doing something completely on their own, unlike typical large aerospace efforts.

The vehicle is so large, flimsy, slow, and awkward, and requires such special weather conditions, that there will not be any big sport arising from it. Still, it is so much fun to fly that I am sure there will be a few hundred things like this around in five years or so, with a lot of people having good times with them. The only practical value is if Henry Kremer keeps putting up his prizes. I doubt that there is any technological fallout of significance because everything that was done here could have been done in 1915 or thereabouts. The only new thing is the particular control system of tilting horizontal surface coupled with wing warp, but soaring birds have been using that for 50 million years.

The greatest value of a program like this is that it gets into new areas, is a stimulus, and gives us a new perspective to look at many other subjects. In this case, it has forced me to look at low-energy surface vehicles, like the streamlined bicycle. Huge, exciting things will be taking place in ultralight aircraft in the next few years. One thinks about that. The devices and techniques for better physical conditioning, which you certainly get involved with in this subject, are very important. Probably the greatest value in human powered aircraft is that there *is* no practical application; it nourishes the spirit, not the stomach.

THE GOSSAMER CONDOR

Peter B. S. Lissaman

Wilbur Wright wrote with a prose style of great lucidity, very different from the jargon of aerospace communication by which our senses have been dulled over the last decade. He said, "It is possible to fly without motors but not without knowledge and skill. This I conceive to be fortunate, for man, by reason of his greater intellect, can more reasonably hope to equal birds in knowledge than to equal nature in the perfection of her machinery."

Two simple equations underlie the Gossamer Condor concept. The first relates to the power required to fly as a function of the scale of the lifting device and the mass lifted. This is an equation of such simplicity that it certainly could be understood by fluid mechanicists, although they may not necessarily be able to derive it! The second relates to the power of animal musculature. When you consider the power of animals, and when you look at that whole range of animals that inhabit this beautiful world in which we live, you find that in proportion to their size, the smaller animals are significantly stronger than the larger ones. The mouse, technically known as *Mus musculis,* is enormously strong and the elephant is very weak, in terms relating to their own size.

Centered in that marvelous spectrum of animals is that most interesting of all animals, *homo sapiens.* If you combine the power required to fly with the power available for flight from animals, as many of us have done, you come up with several interesting results. It can be shown that birds are well capable of flying; they have frequently been observed to do so. On the other hand so are dogs and cats, but they are either unaware of, or unwilling to exercise, this marvelous gift. You find that man seems about the last animal capable of flying, and, up to 23 August 1977, the California condor was the largest flying animal on earth.

If you go back and ask yourself, what is the philosophy of designing a machine like that and how do you do it, the first thing that you recognize is a point made earlier by Fred Culick that configurational details are really not all that important. You can put auxiliary surfaces in the front or the back or the sides on airplanes, you can put engines in the front or back, on the sides, the

---

Dr. Lissaman is vice president of AeroVironment, Inc.

top, or the bottom. Those are really not the things that one debates about. In fact, one can't help remembering the lyrics, "Every hand's a winner and every hand's a loser, and every gambler knows that the secret to survival is knowing what to hold and what to throw away," and that, I think, is the philosophy of design.

We can go back and look at some of the people who tried to make flying machines in the past. The dream to fly by human power is one of humankind's oldest dreams, deeply ensconced in all our legends and all our atavism. We all know the legends of the Greeks; the greatest designer of all time, Leonardo da Vinci, tried for twelve years to make a flying machine; the Indians of the great rocky Mountains have, with many other tribes, traditions of an eagle man, a bird man. Collectively we have been thinking about it, wanting to do it, for a long time. We have not put those two equations that I described together correctly in all cases. I shall return to some of those instances later.

An Italian machine was one of the first. It could not really fly under human power, it was towed to the initial altitude. A designer may ask: what have its designers done? Well, what they did is the following. They were told, you won't need engines any more. Great, they said, so we'll build a kind of high-wing DC-3 and just forget about nacelles. Now I submit that this is a completely linear way of thinking, the wrong way of thinking. Presumably you have your propulsion system close to your engines because that is what it needs to make it go around; and you put your engines on the wings because they are heavy and so that is a good place to put them. But if you have a weightless propulsion system there is absolutely no reason to configure a ship like that.

The German version, a little later, showed more development. Here the designers realized that they had freedom with the propulsion system; they could do something they could not possibly do if they had to have an engine. However, they still thought they had to design an airplane, and they still thought structures and classical aerodynamics, which again are not appropriate for human-powered flying. Whenever you are dealing with something in which the rules are changed and you attempt an innovative solution of it, you should ask yourself, would this solution have worked under the old rules? If it would have, then your answer is probably wrong. The German aircraft, Mufli, was not capable of human-powered flight either.

The Sumpac, a great effort of one of the British teams, University of Southhampton, may illustrate some points. Please do not regard my observations as serious or personal; I am looking back. We made more mistakes than anybody else, and I am simply trying to perceive what seems to be wrong in retrospect. The Sumpac was just a Mufli extended in all directions; what was being done there is something the British do extremely well, which is to carry a given concept to absolutely ridiculous extremes until it becomes so absurd that you obviously have to give it up.

Now you come to the Puffin, another British ship and a magnificent machine. The propeller was in the back. The designers really recognized they could do things that could not be done if a propeller had to be coupled to a steel engine. But they were still thinking airplane and airplane dynamics, and, most important, light aircraft structure. We are told this airframe involved between 3000 and 5000 man-hours in construction—think of it, 3000 or 4000 man-hours in building something before you even try to test that concept! We may well wonder about that. The Puffin crashed early in its first flight program, and it would have been a very big hassle to put it together again.

One of the things that I think one has to do in design, as in life, is to distinguish between one's goals and one's methods—between one's ends and one's means of reaching them. When you think about human-powered flight, you realize there are all sorts of things you can do with the concept. You can talk about it, that's groovy. You can think about it, that's great too. You can attend meetings on it. Fine. You can make calculations. Great. You can write technical papers. Excellent. You can do designs. You can even build human-powered airplanes. But, we didn't want to do any of these things as an end. We weren't interested in that at all. We wanted to *fly* a human powered airplane, and I submit that *there* is the difference. The Gossamer Condor is an airplane in which every piece is there only to make it fly.

I should like to quote a phrase of the great poet-aviator Antoine de Saint-Exupéry (1900-1944), who said, "Perfection is achieved not when there is nothing more to add, but when there is nothing more to take away." I think that is what you see about the Condor. All you really need is a set of wings 30 m long, airframe weighing less than 30 kg, and a propulsion system. It will fly; it will fly just fine. If you are not too sure about yourself and would like to have some positive longitudinal control and maybe some damping, you put in a canard or a foreplane, and that ship flies beautifully. In fact, as Paul remarked, it flies on and on and on. Straight on! The only trouble is it will not turn corners. We had lots of difficulty making that ship turn corners.

We had lots of advice in our project. Some of it we accepted, some we did not. People would come and visit us all the time; the hangar was a total happening, people playing guitars, dogs and cats, everybody there. Lots of private pilots would fly up to visit us and kick the tires or whatever one did to a human-powered airplane. And as you all know, every pilot is an expert in aerodynamics. I was surprised at the virulence with which many of them regarded that control surface in front. It really bothered them. They said, you know, that was not the way God intended it. If God had meant it to be that way he would have put birds' tails where their beaks were, or something like that.

I was interested in how violent the objections were. Ships do it this way, they said, so do fish. You have to turn these things from the back. We really had long talks about that. I finally found a response to them. I said, well, I have no idea what God intended with airplanes or human relations or anything else for that mattter, but I do know that I have ridden horses for thirty years, and it's awfully hard to make them critters turn by pulling on their tails!

So what we did to turn was tilt the foreplane. It was a stroke of great genius, for which as for many of the other ideas, Paul was entirely responsible. I thought he was totally insane, until his logic convinced me he was right. The foreplane is lifting; when it is tilted, it turns the nose. Now the airplane has a very low moment of inertia in yaw, and a very large moment of inertia in roll, particularly when apparent mass is considered. So it turns in yaw very readily and very easily, and then it banks.

That essentially gives you the configuration you need. The only stuff we did on computers was for the airfoils and the propeller. I wrote a program for that and the propeller worked fine. There may have been compensating errors in the program, but you find that you need not taper the propeller, you need not worry about the hub fairings. All you have to do is twist it right and cover it right and it'll work fine. It seemed to us to have an efficiency above 80%, but of course we do not know precisely.

Now, if you know what it is you are going to do, how are you going to skeleton your idea? What structure do you need? What you end up with is a very, very long spar with a mass, which is essentially the pilot and engine hanging underneath, and you do the most natural thing of all, which is to suspend that mass from a series of cables; and that is the basic structure. That is the main wing spar. Of course, when you fly that, when you land, the wing falls off, and so you put on what is called a cabane or strut or king post, and what are called landing wires. Now you have a very, very stiff structure in the plane of the paper. To stop if from wiggling in and out of the plane of the paper, you add what would be called a boomkin and a bowsprit and you install fore and back stays, and now you have a very stiff cruciform-type structure. Your last problem is that the mass is still wiggling, and so you put some more cables in it, and you end up with shrouds, and all the 72 different flying wires on the Condor. But you do end up with an extremely light ship.

The final touch in any grand plan is execution, and here I think Paul's efforts were very close to genius. This issue is, you know what you want to do, you know how you do it, but how on earth are you going to get other people to do it, now that you have decided it is so great? That's the aspect called management, a very subtle and a difficult thing. We have a series of jokes here about meetings. A lot of people think management consists of meetings. Of course it does no such thing. It consists of managing. And meetings are what destroy all our time. Condor meetings were held differently, on very simple principles. There were three: (1) all meetings were held standing up; (2) immediately after a meeting, a decision was made and that decision was executed; and (3) if people concerned with the decision were not present at the meeting, that was their problem. It worked very well.

We have had all sorts of interesting responses to the Condor. Technologists are fascinated; artists adore it. But I see it not as primarily a technological achievement, but as an illustration of the grandeur of the human spirit. I like to think that that greatest poet of all time, whose language we are privileged to share, would be delighted in this manifestation of man's infinite faculty. He's the one who said, "What a piece of work is man! How noble in reason! how infinite in faculty! in form, in moving, how express and admirable! in action how like an angel! in apprehension how like a god! the beauty of the world! the paragon of animals!"

I want to finish by reading you a poem that was sent me by a woman from Africa. The poem is written in syllabic verse:

Here by the Pacific
The blank space of morning
Was filled with the blueprint of their dreams.
They had harnessed the wind
In silent, ornithic horizons
And watched the craft lift off.
Fragile and transparent, shimmering in a million atoms of inspiration
The wings caught the California sunshine
In a triumph of splendor
And fusion of muscle, mind, imagination.
The Gossamer Condor was airborne.
Oh bright was their vision!

STRUCTURAL DEVELOPMENTS FROM KITTY HAWK TO THE GOSSAMER CONDOR

Ernest E. Sechler

It was my very good luck to have been present on the Caltech campus in 1928, when the Guggenheim Aeronautical Laboratory was started. My undergraduate work was in mechanical engineering with an emphasis on design, and I started to work on courses in applied elasticity and aeronautics under one of the best teachers in the world, Theodore von Kármán. Others think of him as an aerodynamicist, or as an expert in fluid mechanics, but to me he will always be the ideal engineering designer. In fact, his first German degree was in civil engineering. Today, I should like to take you on a hurried trip through the structural developments in aircraft since the Wright brothers, with illustrations of the trends in structural configurations and a brief discussion of the underlying structure, and of the design and analysis methods used at that time.

Although I could have developed this whole theme around the aircraft of the United States, I chose not to do so, primarily in order to illustrate the fact that parallel development took place elsewhere, particularly in Britain, France, Germany, and Italy. The simplest flying structure I was able to locate in the literature is an ancient woodcut showing a trainer. The only control was of the canard type, which was copied by many of the early aircraft. If you have been seeing recent pictures of witch flight, you will recall that this control has been relegated to the rear as in almost all other modern aircraft.

Structural development revolves around two major factors: materials and analysis methods. However, underlying everything the aircraft structural engineer does is the fact that the product must be light enough to fly and must be aerodynamically clean enough to have the required performance. Finally, but of great importance since the device will carry one or more humans, it must be safe under all anticipated operating conditions. I have divided the development into six eras. The first, 1903-1912, had wood, wire, cloth, and some relatively poor steel as materials. Analysis, if any, was limited. It consisted largely of first trying to fly the device as a kite, then as a glider, and finally under power, and when something failed under any of these conditions, rebuilding it stronger and trying again.

Dr. Sechler (1905-1979) was professor of aeronautics at Caltech from 1937 to 1976, and executive officer of aeronautics during 1966-1971.

Although the Wright brothers used wind-tunnel tests to determine control characteristics and qualitative effects of lift, drag, and moment, they had no concept of the drag of wire bracing or interference drag of various elements of the structure. The Wright brothers' Flyer of 1903 looked something like a birdcage with all the structure visible. The Blériot Type XI of 1909 was a monoplane in which a first attempt was made to cover the fuselage at least partially. The Curtiss flying boat of 1912 had the first fully covered fuselage. This covering was obviously necessary since the craft had to be watertight. By that time, better knowledge had been obtained concerning the aerodynamic loads and redundant-truss analysis was commonly used. Wings were still thin, and designers—except for a few brave souls—made them biplane in configuration. Monoplanes had to be highly braced to withstand the loads and this bracing contributed greatly to the drag of the machine.

The second era, 1912 to 1918, brought forth configurations that appear more modern. Wood was still used for some designs; for example, the Deperdussin Racer of 1912 had a light wood fuselage and began to look like a modern airplane. The fuselage was fully covered; it was a monoplane, although it still had to be braced, and the wheels and landing gear were streamlined to a limited extent. Others of that era were the Taube of 1914, a German trainer, really a throwback to the previous era, which shows how slowly some concepts develop. The famous French Spad of 1917, and the Gotha G-2 of 1917, the large German bomber of that time, are other examples.

Performance became important during the war of 1914-1918 and more detailed knowledge of aerodynamic loads was sought; the analyses used were still largely truss-analysis techniques carried over from civil engineering. Better grades of steel, particularly as tubing, were becoming available; steel tubing was used extensively for fuselage and engine mounts. Shear was almost universally taken by diagonal drag wires between the longerons and verticals of the fuselage truss. Wood was still being used for many wing constructions and the whole system was covered with dope cloth for aerodynamic smoothness.

In the period 1918 to 1930 more attention was given to the airplane as a means of tranportation and as a private and commercial carrier. Until the early 1930s, structure was still largely fabric-covered steel truss, but more wind-tunnel tests had resulted in thicker wings and a strong tendency toward monoplane configurations. The Fokker tri-motor of 1925 was designed as a transport, and actually was used by Admiral Byrd in his North Pole flight. Its wings were thicker, so that bracing would be reduced to a minimum.

The first inexpensive private plane, the De Havilland Moth of 1925, was the forerunner of many of the private planes of today, although the present craft have become more streamlined; nearly all of them are monoplanes. The pilot was back out of the wind, too. The structure is usually relatively simple and economy in construction is emphasized. Lindbergh's Spirit of St. Louis of 1927, made by Ryan in San Diego, had additional streamlining and metal construction in the engine region. However, this metal was not used as a monocoque engine mount; it was a cover over the basic structure. Wings were braced

Blériot XI. (Courtesy Smithsonian Institution.)

Spad VII. (Courtesy Smithsonian Institution.)

Fokker Tri-motor. (Courtesy Smithsonian Institution.)

by streamlined struts instead of wires, since it had been shown that such struts had less drag. In 1931 came the Piper Cub, a design that has not materially changed to the present day.

Around 1930 a new material made the scene: aluminum, one of the lightest metals. Some of its alloys had strength properties that were attractive to the airplane designer. Sheet metal was available, and under certain alloying and heat treatment conditions the material could be machined, formed, extruded, and cast; it was hailed as the miracle metal of the century. Like all miracle metals, it had hidden problems, intergranular corrosion and fatigue sensitivity in some alloys, but in the course of time these problems were overcome.

Actually, if aluminum alloys were to have been used only for aircraft, their cost would have been prohibitive. The required tonnage was so small that the major supplier, Alcoa, could not have afforded a production line. The aircraft material had to have a very high quality control and precise dimensional and physical properties. However, it was found that aluminum alloys could also be used in large quantities for such mundane things as pots and pans, and civilian structural and ornamental items. According to Templin of Alcoa, these things paid the tab for the development of aircraft aluminum alloys; so in this case, advance in aircraft materials was really a spinoff from other major commercial programs.

In the early 1930s some brave souls, Jack Northrop and Maj Klein among them, thus came to the conclusion that if sheet metal could be properly configured, it could not only form an aerodynamically smooth surface, but could also be used as the primary structure for a vehicle. Research groups in the United States and Europe began to discover that sheet metal, if properly stiffened, could even be used beyond the load at which it buckled—complete violation of the thinking of civil engineers, who had always set the buckling load of any structural element as the ultimate allowable load. Unfortunately, they still do.

Hubert Wagner of Germany showed that thin shear webs could be used for wing spar construction and, if properly stiffened, could carry shear loads far beyond a load at which they went into the wave state. Karl Marguerre in Europe, and Kármán and Sechler in the United States, developed the semi-empirical theory of the effective width, which permits a stiffened flat plate to carry loads far in excess of the load that would cause plate buckling.

Since fuselages and some parts of this wing structure had considerable curvature, a rash of investigators in government and industrial laboratories, as well as in academia, made an effort to determine the load-carrying capabilities of thin-walled cylindrical structures, as well as other sheet construction with a reasonable radius of curvature. Relatively quickly empirical design curves were established that are still in general use. Only during the past few years have Charles D. Babcock and Johann Arbocz of GALCIT finally determined the major parameter leading to failure of thin-walled cylindrical structures.

Since failure in such thin curved sheets tended to be catastrophic, the design curves used were lower-bound values of all experimental tests; we thus had a built-in factor of safety which, during the period 1930 to the middle 1960s, has made this type of structure very forgiving of unexpected loads and structural damage. The Sikorsky S-42 was one of the Pan American Clippers of 1934. At that time it was proposed that the ocean would make a better and cheaper landing field than one on land and, at least for overseas flights, there would be a built-in safety factor in having a boat along in case of engine failure. It soon became obvious that the flying boat's disadvantages were greater than its advantages, and all major efforts were later concentrated on land-based aircraft. Actually, one of the disadvantages *was* in the landing field. A harbor is one of the dirtiest things in the world and a landing boat coming in and hitting a log is not too good. That was one of the reasons they had some problems.

The Douglas DC-3 of 1935 was the forerunner of all future commercial aircraft configurations. It was of semimonocoque stiffened-thin-sheet construction throughout; the skin was not only used to form the aerodynamic shape but it was also the primary load-carrying structure. Because the wing skin was load carrying, it was strong enough and stiff enough to eliminate all external bracing, which made for a very clean aerodynamic shape. Low-drag engine cowlings had been developed. Except for the nonretractable landing gear and minor external antennas, there were no external items to increase drag, and even they were later eliminated by the retractable landing gear and the flush antenna.

During 1935 to 1949, which included World War II, the major effort was in the development of military aircraft. For cargo and troop transports, modifications of the DC series were widely used. The Boeing Flying Fortress was first flown in 1935. During military action, these aircraft proved the value of semimonocoque construction, since many of them suffered severe combat damage but were still able to return to their bases. Other World War II airplanes were the British Hurricane of 1936 and the German Heinkel 176 of 1939, the first jet airplane. Fortunately, Hitler did not believe in it and production was stopped; otherwise it would have been one of the German aircraft we would have had to face.

An unusual airplane was the De Havilland Mosquito of 1942. This versatile aircraft was constructed of laminated wood and was used for almost all types of combat missions. Because of its success, considerable pressure developed in the United States to use wood for aircraft construction, to relieve the demand on aluminum, which was a strategic material. However, serious study showed that wood of aircraft quality was an even scarcer material than aluminum, and after a few partially successful attempts in America and in Europe, wood was abandoned as a major structural material, at least for large aircraft. Everybody thought that plywood was cheap and easy to get; but as to aircraft-quality wood at that time, there just wasn't enough of it in the world.

Sikorsky S-42. (Courtesy Smithsonian Institution.)

De Havilland DH-98 'Mosquito.' (Courtesy Smithsonian Institution.)

The Messerschmitt 262 of 1942 was the first operational jet; the North American F-86 Saber of 1947 was a postwar version.

During this period the major new problem faced by the structural designer, aside from the high additional load factors for military aircraft, was the introduction of the pressurized fuselage. This feature was introduced so that the plane could fly higher. It was advertised to the public as enabling aircraft to get above weather disturbances, without decreasing passenger comfort. Therefore, the fuselage not only had to carry the flight and landing loads, but also an internal pressure that added considerably to the stress on the skin. One relieving factor was that internal pressure increased the maximum buckling load of the skin when the skin was subjected to compression. On the other hand, there was the very serious additional problem of possible explosive decompression in case of skin rupture. One only has to figure the amount of energy stored in an airplane with 8 psi pressure to understand that one has got a bomb on one's hands. The British Comet of 1949 had the misfortune of being the first to fail in this manner, by fatigue crack propagation and a stress concentration radian due to cyclic pressure loading. This failure led to a flurry of tests and analyses, which culminated in the fail-safe concept for all transports and many other aircraft. According to this concept, if a structural failure occurs it must not progress catastrophically to complete failure of any major structural element. Thus, a crack in the fuselage should not lead to an explosion, but to a leak that can be detected and repaired for future operation.

Up to the present day, the basic structural materials have been high-strength aluminum alloys for the major portion of the structure, high-strength steels for landing gear, and titanium alloys for areas around the engine subjected to high temperatures. Development of new machining techniques has made it possible to machine stiffeners and reinforcements integrably with the skin, so that jigging and riveting together of large numbers of small parts are eliminated. This advance has led to significant economies in production.

Analysis methods have become more and more refined, primarily owing to the development of high-speed computers. All analysis was previously carried out on desk calculators; handling a 20 × 20 matrix was a difficult and time-consuming task. Modern computers treat matrices two orders of magnitude greater than that and do so in a small fraction of the former time. This advance has greatly increased the analyst's ability to approach an optimum structure—particularly in the dynamics area—to investigate modes, frequencies, and stresses that could not be studied by hand-calculation methods. In fact, there is some concern that some designers and analysts have become so enamored of the big computer that they have lost sight of the fact that the designer is supposed to think and to come up with innovative structural concepts that have not already been programmed into the computer.

The current state of the art is exemplified by the Douglas DC-10, the Lockheed L-1011, and the Boeing 747. Although they represent the larger wide-bodied aircraft, smaller versions look very much like them and have the same basic structure constructed from the same materials. So far, except for some of the military aircraft, we have been discussing subsonic or nearly transonic vehicles. A little later we shall take a short look at super- and hypersonic possibilities.

De Havilland 'Comet' commercial jet transport. (Courtesy Smithsonian Institution.)

During the past ten years a new material concept has become more and more interesting, the so-called composite materials consisting of fibers of very high strength and high stiffness embedded in an organic or metallic matrix. Whereas the common metals, such as aluminum and titanium alloys and steels, all have about the same strength-to-weight and stiffness-to-weight ratios, some of the composite materials have ratios that are much higher. They therefore promise greater strength or greater stiffness for the same weight of structure, or conversely, a lighter structure for a given strength or stiffness. Since the material is made up of a combination of fibers, the orientation of the fiber determines the directional properties of the final material. Thus, the material can be tailored to specifically oriented material properties which could tend toward an optimum design for any specified loading condition. Another advantage is that use of the proper types of forms in which they are manufactured and cured, complex structures with compound curvatures can be constructed with integral stiffeners and reinforcements. Additional parts or large assemblies can be joined by a cementing process rather than by riveting or other individual fasteners.

Despite their many potential advantages, composite structures must still be used with proper respect for their disadvantages. The resin matrix materials with graphite or boron fibers are brittle; care must be used to avoid stress concentrations, regions where cracks could start. Their properties after long exposure to various environments are still somewhat largely unknown. Their failure mechanism is very complex and difficult to predict. Furthermore, their material properties are very sensitive to many manufacturing parameters that are difficult to control to the required degree of accuracy. Quality control is therefore in many cases poor, and advantage can seldom be taken of their full potential.

Nevertheless, I see a great increase in the use of composite materials in all phases of aircraft construction, in the near future, but I think some form of order will have to be established in their manufacture and utilization, both in the organic and in the metallic. The organic are with us now; the metallic composites of fibers and metals are still being worked on as a research tool.

Costs must be reduced, which means we shall need quantity production. We shall have mass production, at least of the simpler form, such as sheet or tubing. There is no particular reason every company should build its own sheet stock and roll its own tubing merely because most of the companies have not enough quantity production to set up proper quality control. Tighter quality control is essential, but I think that we should leave the detailed analysis of failure mechanisms to research groups and use the material just like we use aluminum.

At present we are doing something we never did with the aluminum alloys. Aircraft companies never tried to devise new aluminum alloys—they left that to the major aluminum companies. Now they *are* trying to build up composite materials, but in many cases they lack the quality control or the manufacturing methods to do it correctly. Also, design construction and analysis methods must be formalized, at least to a reasonable extent. At present everybody is developing a new learning curve each time the use of a composite material is contemplated. That costs money and takes time. This is a new material and new design methods must be used. It is not a one-to-one replacement for a metallic structure.

As to some of the future of analysis, the complex computer programs such as NASTRAN and STAGS, and many others, will be used continually. They are beginning to be used with more and more understanding and respect for what they could do for final analysis. I see the need for a simpler set of programs for preliminary design, for trend studies that do not take into account detailed analysis techniques but actually bring to the designer various configurations and possible structural configurations to handle that particular problem—something that can be pulled out of the computer, looked at, studied, put together as a patchwork, and then finally used in a more detailed design. It will probably be an extension of the current rather small effort in computer-aided design, but it should pay off significantly in the early design phases of any new aircraft concept.

Rather than leaving you with the feeling that all the fun has gone out of the structural part of the aircraft industry, and that all that is left to be done is to polish up old concepts, I should like to take just a minute or two to look at where we might be going. Take the British-French Concorde of 1976, the only supersonic airplane in commercial operation. This Mach 2 airplane probably represents the limit in the use of aluminum alloys, since above that speed they cannot handle the increased temperatures, at least not in a simple manner. Is there any reason to think we may want to go faster?

Well, history has shown that all new major steps in aviation have been taken by the military. If the Air Force builds a faster and a higher-flying airplane, sooner or later the civilian population will want to fly just as high and just as fast. We already have the McDonnell-Douglas F-15 of 1972, which flies over 1700 mph, and the Lockheed SR-71, which flies at a very high altitude and at Mach 3-plus speeds. But these airplanes are not fast enough for some military missions and therefore studies are currently under way for aircraft in the hypersonic range, which means Mach 5 to 8 or even higher speeds.

As a sideline to these military aircraft studies, most of the major aircraft companies have small teams looking into the possibility of hypersonic transports. I am sure they will eventually be developed and will be flown, and people will ride in them. These aircraft will call for completely new structural concepts. Insulation and cooling will be major problems, and if hydrogen is to be used as fuel, which is a real possibility, whole new ranges of temperature will have to be handled by the structural engineer. The structural challenge is essentially here today. It is a challenge that will be fun to work on, and I am confident that future members of the GALCIT family will make important contributions to these new developments.

THE SKUNK WORKS

C. L. (Kelly) Johnson

No one told me what kind of a meeting this was going to be today, and I came here first because I thought it was just a reunion. Nor did anyone tell me that I was going to be upstaged by a Skunk Works that built the Goosamer *(sic)* Condor. But without even my 10in. Michigan slide rule I figured out this noon that while they are flying that airplane with the equivalent of the output of one man, with just 1.7 million men we could get the performance that we are getting out of the SR-71.

Before I start to discuss the Skunk Works, I should like to give you a quote from Larry Kitchen, who made the following statement in a recent speech:

> We are lucky the airplane was invented back in 1903 and developed over the following years. We would probably be stopped in developing the airplane in today's climate. The environmentalists would object to the noise, the consumers would want to have them all recalled before they got out the factory door, the pacifists would complain that they could be used to carry troops, preservationists would be afraid they could fly into the city hall, and ten government study groups would prove that they definitely cause cancer.

Well, I am glad I have not run into those restrictions during my tenure at the Skunk Works or at Lockheed up until recently. We have been very fortunate in the types of contracts and operations that the government has allowed us to use in the Skunk Works.

First, how did the Skunk Works start? For about four years prior to 1943 we at Lockheed had been trying to get permission from the Air Force to build a jet engine and a jet airplane. We worked all the way thorough the war trying to make the P-38 gain 17 mph, but between the compressibility effect on the propeller and on the wing, that was all we got after we doubled the power. So, it was obvious something had to be done. Finally, General Frank Carroll said, All right, Kelly, we'll give you a contract but we'll give you one for the engine you're proposing and one for the airplane you're proposing, but you've got to use a British-designed engine; how soon can you have it flying? The pressure at that time was to get something to fight the ME-262, which was available in Germany in large numbers, except that we were keeping them from getting their fuel. I said, Well, six months after we get a go-ahead, we'll be flying.

Mr. Johnson is retired as director of Lockheed's Advanced Development Projects Group and is now consultant. He has been at Lockheed since 1933.

When will the time start? He said, it'll start this afternoon; there's a plane back at two o'clock and you'll have a letter of intent. We got the letter of intent, and nine days later we had a conference on the mock-up design, which was the original P-80.

We built the first Skunk Works using our wind tunnel model shop in Burbank, some Wright engine boxes left over from the Hudson bomber days, and a rented circus tent. In 143 days we rolled out the first airplane, took it up to Edwards, and flew it. Unfortunately, we had gone so fast that the engine contracts had not yet been let to Allis-Chalmers to make the engines. They could not execute the contract by the time we got the airplane flying. So they told us, all right, you'll have to do it over again, but this time you'll use a General Electric engine, the I-40. And so in 139 days we built an airplane 30% bigger, that went on to become the F-80A and its successors.

How did the name Skunk Works come about? One of the first things we did and have done ever since, was to clamp strict security on our operating area, not only engineering, but the shop, purchasing, and everything else. Someone asked one day, What's Kelly doing in there? And the answer was, Oh, he's in there stirring up some kind of a brew. Well, those of you familiar with Al Capp's *L'il Abner* cartoon strip in the funny papers might recall seeing Hairy Joe and the Indian stirring up Kickapoo Joy Juice and throwing in a skunk, and anything they had handy, to make this a very strong liquor. So when they came down to the fact that we were stirring up a juice, someone said, Oh yes, it must be the Skunk Works. Well, just nine years ago we copyrighted the name; it had become well enough known that way, and most people do not

P-38 Lightning (1939).

know us as the Advance Development Projects Group of Lockheed, they know us as the Skunk Works. And that is how it started.

We have operated continuously since 23 June 1943, starting out as an engineering experimental division with the F-80 and its derivatives. For a long time we would make only two of any kind of airplane, do the development work, and then turn production over to one of the other Lockheed divisions. On the F-80, it was to the California division in Burbank; the F-104 also went to Burbank, the C-130 went to Georgia, and the other airplanes were distributed until we got to the U-2. There, the security classification was so high we decided we would have to go into production ourselves, so we became not only an experimental division but a production facility working up to a production of almost five airplanes a month.

Then when the SR-71 and the Blackbird and its derivatives came along, we again had a very tight security situation. The Skunk Works, at the peak of its operations, grew to a total of over 10 300 people in Burbank alone. In dollar value, the maximum output was then the equivalent to a production of sixteen Constellations a month. So we had come a long way from making one or two of something.

I want to mention some of the airplanes we made and give you my own evaluation of whether they are good, bad, or indifferent. As mentioned, the Skunk Works started with the F-80 and the F-80A. The ratio in weight between the first one and the one that finally went into production was 1:2. That was an excellent program. From the F-80A came the T-33 trainer, of which we made about 7000. That worked fine. And then there were several interceptors.

Then we came to our first real failure, the XFV-1. That was a tail-sitting, vertical-rising Navy airplane designed to land on the deck of a pitching carrier, or destroyer, going up and down 17 ft while backing down with the pilot looking over his shoulder. Well, our cost performance was good. We did it for the cost. We practiced on clouds, but it got to the point where we decided the pilot could not look backward over his shoulder, and judge height accurately enough to land the thing. Besides, the power plant had a funny habit of having its own ideas of what to do in the low-power region: it would change thrust 25% at the critical time. I had the job of writing to the Navy and saying, Dear Navy, this is the first airplane we have built that we're afraid to fly ourselves. Whereupon it was cancelled without too much expense.

The Saturn was a small airplane for feeder airlines and turned out to be a pretty good airplane. We built two of them. But we managed to have a very poor market situation in that when the Saturn came out for a selling price of $75 000 apiece, there were approximately 9000 C-47s on sale after the war. Whereupon we drove a tractor over them (over the Saturns, that is, not the C-47s; they are still around).

The Constitution used our Skunk Works practices and engineering. It was a very large airplane, the first widebodied transport if you please, and double-decked at that. We built two of them at 184 000 lb gross weight, but it was designed for turboprops that we never got, and it was the world's most underpowered airplane except the Gossamer Condor. But it had similar take-off characteristics. The Navy ran them until they ran out of spares and then one of them became a hotel or cafe up in Las Vegas and the other one was planted over somewhere in Arizona and destroyed there just this past year. A fine airplane, except it was underpowered.

The XF-90 was the world's strongest airplane, the only one ever to defeat the atom bomb. It was designed on the basis of things we learned in World War II and it was just hell for strong. It was designed for 14 load factors to fly fast at low altitude. When they put it up at Frenchman Flats along with some other airplanes and set off an atom bomb near it, it cracked the canopy. There's a case of a poor specification, looking backward.

Then came the X-7, which is a ramjet test vehicle and that we managed to fly, the Marquardt ramjet in it, to speeds up to Mach 4 at 95 000 ft. We learned a lot from that, in terms of the straight thin wing, that we were able to apply to the F-104. Then we came to the R7V1, which is putting turbo props on Constellations. It was a dandy airplane and in level flight it could outspeed a P-38 in a dive, but it would only fly half way across the Atlantic because the fuel consumption was so bad. We built four of them, two for the Air Force and two for the Navy. They were run for a while and I do not know what has happened to them.

Then came the XF-104. We built two of them in the Skunk Works and our performance on that was excellent. We ended up eventually building some 2500 around the world, many still in service. Then came the YC-130, which we built too, and turned over to Georgia. That aircraft turned out very well and we were able to hand it off very nicely to the Georgia division. Then came the U-2, which we built in eight months to fly, under price, on time; we built quite a few of them. The performance of that one was very good and in fact we shall probably reinstitute the line on advanced versions of that airplane, known as the TR-1. We built the first two Jet Stars.

Then we came to our very advanced airplane, the CL-400, which was a liquid-hydrogen airplane that would cruise at Mach 2-3 at around 100 000 ft. We learned how to handle liquid hydrogen, but the thing that killed it was the problem of getting liquid hydrogen around to the aircraft wherever it was based. Had we gone forward with that program in 1972, we would have used 10% of the total natural gas coming into Los Angeles just to make liquid hydrogen. We would have really hit it right, wouldn't we? So that one went down the drain.

The Model A was the first of the Blackbird airplanes, known as Oxcart, and that was the first Mach 3-plus airplane to cruise between 80 000 and 90 000 ft. From that came the YF-12A interceptor, and then the SR-71. The SR-71 has been in service with the U.S. Air Force since 1973. It has made over 15 000 aerial refuellings. That was the one on which we reached the highest production and the one which is probably the most advanced reconaissance aircraft in the world today, as far as we know. Then there are two other programs, both secret, and doing well.

Now one of the reasons we were able to get various government contracts, in general, has been that we have made unsolicited proposals for very low costs and promising very fast development time to first flight. The XP-80 was started in 1943 and 1944, 1140 days; we had a string of them up to the F-104, which took a whole year. The U-2 and the Jet Star were 8 months apiece; the YF-12 was 36 months, the first titanium airplane, and that was a tough one. The SR-71, which was really a derivative of the prior ones, took two years. The U-2R, which was a brand new U-2, took a year.

As to the conditions under which we operate, we always try to make them contractual conditions for meeting our costs and time of performance. Here are some of our basic operating rules, and unless we follow them, we cannot do any better than anybody else.

First, the Skunk Works manager must be delegated almost complete control of his program in all aspects. He should report to a division president or higher. That has been a cornerstone of our whole approach; I have been given authority directly under the chairman of the board or the president of the company, to have a self-contained division of Lockheed. We locked out everybody. Unfortunately, one day we locked out the chairman of the board, whereupon one rule was changed. But most companies do not do that. They like to have their committee meetings, and I am death on committees.

Second, strong but small project offices must be provided both by the military and by industry. For instance, the size of the Air Force project office on

U-2.

the SR-71, which cost over a third of a billion dollars for its first contract, was 32 people, including the secretaries. It is quite unusual for the Air Force to have such a small project office on such a large program. And I should add that our project office had six.

The number of people that have any connection with the project must be restricted in an almost vicious manner. Use a small number of good people, from 10 to 25% of what is used in the so-called normal systems. In engineering it is 10%, in the shop it is 25% because we lean heavily on our ability to use drawings directly, and we need heavy supervision and inspection to make sure the parts come out right. We use a very simple drawing and drawing release system with great flexibility for making changes. One of the rules is that we put the engineers right next to the hardware, and the person building it or tooling it or purchasing for it has no excuse not to know what the job is. In fact, 600 ft from the engineering to the person who uses the drawing is a long way yet quite unusual even in other divisions at Lockheed, where it is anywhere from one to two miles, and drawings never do get to the person who makes the parts. They go to a planning department. And then I don't know what happens.

A minimum number of reports must be required but important work must be recorded thoroughly. I do not like the situation that existed at the start of the Blackbird program, where the flight-control system reporting, for instance on the Apollo, was 200 000 8½ × 11 pages a month. On the Dinosaur, it was something like 6000 pages a month, and the same company was doing the same work and they started to send me progress reports. The first one had 330 pages and in the middle of it I ran across a proof of Bernoulli's theorem, which I used to teach. Whereupon the monthly progress report was kept to a maximum of 30 pages over seven years. If you cannot explain what you have done in a month in 30 pages, you don't know how to write, which is not strange.

There must be a monthly cost review. I won't go into that, but it is very important. The books are no longer ninety days late, because nowadays a computer can give you a report every morning, if you have put the right things in. The contractor must be given and must assume more than normal responsibility to get good vendor bids for subcontract work. Commercial bid procedures are often better than military ones. The systems we use to buy parts or transports are tougher than the military ones. We get better prices and better delivery. But it is still easier and less paperwork than the military ones. I won't go into the inspection system, but it is a very important part of the whole thing.

Another rule, which always makes my military friends mad, is that the contractor must be given the authority to test his final product in flight. He can and must test it in the initial stages. If he does not, he rapidly loses his competence to design other vehicles. The trend is increasingly for the military to take over all the flying on new aircraft and missiles. In missiles, the trend was established long ago and I guess we won't change that one. But in aircraft, if

SR-71.

SR-71.

the manufacturer does not get hell scared out of him a few times a year, he will lose his competence to make any more airplanes. So we make it a contractual requirement for us and it is an uphill battle every year.

The specifications that apply to the hardware must be agreed to in advance of the contracting, and it is our practice to have a specification section that clearly states with which important military specifications items we know we cannot comply and why. That is a good practice and saves all kinds of arguments. We tell them not only what they are going to get in our normal spec, but also what we are not going to do. And we try to settle in advance before we sign a contract; in fact we get an agreement with it as to what we are not going to do. In an unusual aircraft like the SR-71 and others, there is very little in the handbooks and in the specifications that can be applied right across the board. The list of exceptions is bigger than the list of what you are going to do.

Funding a program must be timely. Only twice in twenty-four programs has Lockheed had to carry on and support government programs. There must be mutual trust between the military project organization and the contractor, close cooperation liaison on a day-to-day basis. It used to be, before telephones were tapped so much, that our project office was as close as the green button on my desk. Now it is a little more difficult than that, but there are ways of communicating with proper security, and I claim that we keep our project office informed better with our direct contact than they would under all of the normal methods of approach that have grown so much like Topsy.

I mentioned access by outsiders to the project. I love security. It protects us from being overhelped by Lockheed. If they're not cleared, they can't come in—and we don't clear them. And here is one that took seven years to get from Lockheed. Because only a few people will be used in engineering and most other areas, ways must be provided to reward good performance by pay based not on the number of personnel supervised. And that attracts good people and keeps the ones you've got.

Well, if the Skunk Works is so successful as some people say it is, why doesn't everybody do it, not even at other divisions at Lockheed? Well, here's the one that almost gets me fired. I believe industrial management does not want to use our type of operation because fewer people are required; therefore profit is less, because profit is a function of the number of people on the program. Too bad, but there it is. It requires almost total delegation of wide power to a very few individuals. Neither industry nor government like that too well. It provides no place for committee action. Decisions must be made at 7:00 a.m. every day except when we start early, when they must be made before then. Only very good people must be assigned to the project.

Well, one thing I found out very early when Mr. Gross said, O.K., Kelly, of course you can do these things in a hurry; but look, you've got the best people in the place. I said, Mr. Gross, we have a method of rating people in engineering on their performance, based from 1 to 5; I'll do the next program with almost all 2s and 3s. We did, and lo and behold, when the people were given responsibility, told what the job was and put in contact with the people who were building it, they rose to it and it was not very long before they were rated as 1s in performance. They upgrade themselves and love it. So I told Mr. Gross, he had enough people around to staff at least six Skunk Works, if he would delegate the authority. That is the last I heard from him. It tears down many long-established empires.

The hardest one to get was our self-contained purchasing department. We did; and the Skunk Works is probably the most audited outfit in the whole country, not only by everyone from the Office of Management and Budget down, but by our own corporation and companies. Finally I suggested we do something else. If we are to have contracts as large as these, let the military put an auditor who would live right above my office, and then before we took action on some of these things and went sole-source or did things that were not according to the Air Force System Procurement Regulations, I would tell him what we were going to do. Then I would write a piece of paper and put it in a file, and unless he told me within a week that I should or should not do it, that was the end of it until Senator Proxmire came around with his Golden Fleece award. And then I have a whole file that says, Look, we told you what we were doing, this is what it was, and we have been audited—what else do you want? A good system.

Our methods are hard on the key people, both military and civilian. They have no large groups of people to supervise, and it often costs them promotions and pay, particularly when high security is involved. On some programs the military officer's direct boss and the people who wrote his efficiency rating did not know what he was doing. And if I try to say, Colonel so-and-so is doing a fine job, they said, Oh phooey, Kelly is just trying to scratch our back. So the poor fellows would be working there for years doing a magnificent job and could not talk about it. Now that is not so true in the civilian area, because we can talk and get enough salary and so on.

If production is to be started by an organization different from the prototype group, the N.I.H. (not invented here) factor can result in excessive changes, delays, and increased costs. It is extremely difficult to hand over something that you have built and tested, and say, here, make more like this. It never happens. They have to change it. They changed $92 million worth on the Jet Star. Lost our shirt. Built the first airplane with something like 55 people in engineering and 200 in the shop and they had 535 in engineering to redesign it. So you have got to watch that N.I.H. thing.

We have talked a great deal about prototyping and here are my comments on that. First, I am totally dedicated to prototyping when management methods are tailored to the degree of risk. There is only one way to do the best for each project, from a Gossamer Condor to the C-5. There is no one best way that will apply to all of them. Second, we prototype things we expect to produce, otherwise it is just fun for the engineers and a heartache for the taxpayer. I object very strenuously to the method taken to develop the lightweight fighter. I thought we should go directly to a design that could be produced because it was not reaching so far into new technology that we would have to go through two stages. We are not doing that and we have lost the advantage of the lightweight fighters going the prototype route plus the normal route for production. In fact, we are going through two prototype programs, at a cost that is terrific, and a cost in time of about 2-3 years. I do not make a lot of friends with Convair, the Air Force, or the Department of Defense saying that, but that is the fact. Third, consideration is given to production during the prototype phase. This can be done very cheaply (about 2 to 4% of the prototype costs), and can sometimes cut production costs substantially. When we did the experimental drawings for the F-104, we released them not only to our shop, but I had twenty engineers separate from the basic engineering group, whose directive was to find out the cheapest, best way to make the part with no compromise of weight, accessibility or maintainability, or drag. And it was a very successful program. For something like $400 000 we saved millions and millions of dollars when we went into production. And we were immediately in a position to go to production with things that we could not do on the prototype airplane—cast inlet ducts and all kinds of other things. Yet we did not do it on the lightweight fighter, so there are probably very few pieces of prototype that actually look like the production parts.

There is a clear definition and a separation of the responsibility and authority of each party, military and industrial. That is a hard one to get. We have been fortunate, and as long as you are successful, I guess they will give you this sometimes. So that is the basis of how the Skunk Works operates, and we have been through many, many programs. I do not know what the number of the one would be right now, but we are still considered as a Skunk Works, and holding the line, holding off both Lockheed and D.O.D. quite well, at this point. And we are going to continue.

In conclusion, let me say something about the development of the SR-71, which was a very difficult airplane to build. In terms of metallurgy, little information was available on titanium. We were building an airplane the metal of which had operating temperatures between 800° and 900°F. We would have to seal the fuel tanks against an average wing temperature of 560°F, and needed wiring that would operate in a nacelle at an average temperature of the air flowing by it of 1500°F. So we had to invent everything from tank sealant to a type of wiring that would take that, to thirteen different designs of an oil temperature transducer, to tires that would take the temperatures of prolonged soaking at the high temperatures we got in cruise conditions. We had to cruise not for two or three hours in afterburning but for periods of up to ten hours, with refuelling of course. We used a turbo ramjet engine developed by Pratt & Whitney, the most advanced in the world to this date as far as I am concerned, which operates as a turbojet on up to a Mach 2.8 and then it shifts over into a ramjet cycle, and then it goes to work. It goes fast. And we had to keep it all working.

THE BEGINNING OF ROCKETRY AND JPL

Frank J. Malina

The tale I have to tell involves one of the most extraordinary developments that has occurred in the history of mankind. The practical implications of astronautics as they are now known are fairly clear; I think the philosophical implications are still not at all clear. In this connection I am rather struck sometimes when I hear people say they believe that Caltech is an offspring of the Jet Propulsion Laboratory. I have a difficult time explaining to them that it is not so.

I shall restrict my tale to the period between about 1935 and 1939, when we received the first contract for the support of rocket research from the U.S. government, amounting to the grand sum of $1000. This sum was to be used to study the literature and to prepare a proposal for the possible development of liquid and solid propellant rocket engines for use as auxiliary power plants on aircraft.

In the political and social scene, 1935 was one of the difficult years of the Great Depression. A graduate student at that time working on the wind tunnel was paid about 25¢ an hour. Roosevelt was President. Numerous government agencies were trying to help people survive economically and one could even obtain some funds from the federal Works Progress Administration to pay student labor on research projects. Many feared the threat of war in Europe, and many of us lived through a decade of preparation for and involvement in the events of World War II.

On the scientific and technological scene, in physics, cosmic rays were a puzzle. Physicists were beginning to take an interest in releasing energy from the atom. Chemists had developed many new chemicals and understood reactions between many of them. Combustion was not understood, and I am not sure it is yet. New materials were becoming available for structures: aluminum alloys, strong steels, and so forth; new refractory materials were being developed. Within a few feet of GALCIT, the 200in. mirror of the Palomar telescope was being ground. In the wind tunnel, tests were being conducted on the DC-3, DC-4, and other aircraft. NACA published a report concluding that thermal

---

Dr. Malina (1912-1981) was co-founder and first director of the Jet Propulsion Laboratory, and later founder and editor of the art journal *Leonardo*.

jet propulsion for aircraft was not a practical possibility; unfortunately, Edgar Buckingham's analysis assumed that instead of air compressors of the type now used they would be of a piston type. The theory of aircraft control and stability was quite well advanced. Commercial aviation was becoming a big business and much progress had been made in the development of lightweight aircraft structures.

At GALCIT, the faculty consisted of Theodore von Kármán as director and Clark B. Millikan, Ernest E. Sechler, and A. L. (Maj) Klein. There were also visiting lecturers, primarily from the aircraft industry in southern California. Research ranged from aircraft design to aluminum alloy structures, to study of turbulence, characteristics of propellers, and theoretical studies of bodies in supersonic flow. No work was being done at GALCIT on aircraft power plants.

What was the astronautical scene? What is called the first generation of astronautical pioneers had by 1935 published quite a number of papers and books. There were the studies of K. E. Tsiolkovsky in Russia, R. H. Goddard in the United States, Robert Esnault-Pelterie in France, and Hermann Oberth in Germany. These studies were regarded by "serious scientists" as mostly science fiction, and it is true that they were highly speculative in terms of the state of the art of rocket propulsion at this time. Goddard had a grant from the Guggenheim Foundation and was working in New Mexico in connection with the Smithsonian Institution. The American Rocket Society already existed, there was the British Interplanetary Society in Britain, and a number of others formed groups to dream of escaping from the earth. In Europe, Eugen Sänger was working in Vienna and had published a paper on some tests he had been making on a rocket motor. This paper was reviewed by William Bollay at a GALCIT seminar in 1935; I shall point out the consequences of this review later.

In the 1930s in Germany, Walter Dornberger, Oberth, and Wernher von Braun had begun work for the Nazi government and their work was secret, so we did not know anything about it until about 1943. There was also work going on in the Soviet Union; we have only learned about that in the past few years.

I came to Caltech from Texas A&M on a scholarship in mechanical engineering in 1934. By the third term I had a scholarship in both mechanical and aeronautical engineering, and had started working on the wind tunnel. With much interest I watched A. E. Russell and H. M. McCoy, who later became Admiral Russell and General McCoy, studying the characteristics of propellers with a power model in the wind tunnel. I offered my help, free of charge, and they accepted. As a result, William Jenny and I wrote a thesis on the characteristics of braked, rocket, and free-wheeling propellers.

It was in March 1935 that Bill Bollay gave a review of Sänger's work in Vienna. Bollay's conclusions were not at that time very optimistic as far as rocket-powered aircraft were concerned, and we have no such aircraft actually speeding their way around the earth at the present time.

This seminar attracted the attention of two young fellows, John W. Parsons and Edward S. Forman. Parsons was a self-trained chemist, rather exotic, of the cultist type, which is of course not unusual in California, and Forman was a skilled mechanic who had been working with Parsons in an attempt to build solid propellant rockets. At that time they were working for an explosives company in the Mojave Desert. They came to GALCIT and said that they wanted someone to advise them on the construction of a liquid-propellant rocket engine. They were sent to me, and I said, Let's see if we can't make some kind of a proposal and see if we can carry out some research at GALCIT. I discussed the possibilities further with Bollay, Parsons, and Forman and drew up a proposed program in about February 1936.

Then I went to Kármán and he said, Well, I think that's quite interesting. You can do it, but no money. We were not discouraged by that and began some experimental work with money from our own pockets, including an experiment that was carried out in the Arroyo Seco in Pasadena in October 1936, with gaseous oxygen and alcohol in a small engine that was cooled with water placed in a jacket around it. This became the first test of a rocket engine at Caltech. There is a replica of the engine and test stand constructed by Walter Powell on view at the museum at JPL, if anyone wants to see the "fancy" equipment we used.

This apparatus was transportable, which was not very convenient, as you can imagine—going out each weekend with cylinders of oxygen and apparatus, setting it up in the Arroyo, and taking it all back to the campus. I asked Kármán whether it would be all right if we made some small-scale experiments inside GALCIT. I explained that we would make very small experiments. So, we designed a new engine; the nozzle had a diameter at the throat of about 1/8 in. As an oxidizer Parsons suggested using nitrogen tetroxide, which is still being used today, and alcohol. We mounted the engine with tankage on a bob of a 50ft pendulum hung from the top floor of GALCIT into what used to be the pump laboratory in the basement.

When A. M. O. Smith and Hsue-shen Tsien heard that we had got approval from Kármán to do some research, they asked if they could join us, so that by the time the experiment was made in the autumn of 1936 there were five of us. When we had set up the pendulum apparatus in the Guggenheim building, one Saturday morning Amo and I went to the Gates chemistry building and obtained a cylinder of N_2O_4 and put it on the lawn in front of Gates; we intended to get about a liter of it. But it was an extremely old, rusted cylinder and the valve jammed. There was a tremendous spout of N_2O_4 onto the lawn; the burnt spot persisted for weeks, much to the chagrin of the gardener.

We carried on with our experiment and, not unusual in rocket research, we had a misfire. Out of our little engine poured a tremendous red cloud of N_2O_4, mixed with alcohol, into the bottom of the Guggenheim building, which seeped upward throughout the building. We did not realize just what the consequences were until we came back on Monday morning. Bill Bowen and other people were waiting for us, and each had a bucket with oily rags. We had to go through the whole building and wipe off the rust that had formed on anything that was made of steel. And Kármán said, Out! We then hung that pendulum on the east end of the building and carried out a few experiments that gave us a bit more experience.

Next we built a "gas apparatus" that provided gaseous oxygen and ethylene to an engine that delivered about 5 lb of thrust. At first we used a carbon nozzle from an electrode we had bought for a small sum, and finally shifted to a copper nozzle upon the urging of Amo Smith. With this apparatus we began measuring the characteristics of exhaust nozzles.

One day in 1939 I was called by Kármán's secretary to take a typewriter to his home, and when I came back I noticed a large number of people standing around GALCIT. As I walked closer I began to see on the ground bits of equipment that I suddenly recognized as part of our gas apparatus. It had blown up. I think I am very fortunate to be here, because if I had been there when the explosion occurred, it is not very likely that I would be here, since a piece of a pressure gauge had passed where my head would have been and buried itself very deeply in a board. This apparatus was rebuilt and later transported to the Arroyo, where we finally set up an "establishment."

When we started this program we had very strong moral support from Robert A. Millikan. He was an extremely open-minded, adventurous type of person. Also, when Irv Krick, who was teaching meteorology and whose group was developing a radiosonde weather instrument to be carried by balloons, heard that we wanted to build a high-altitude sounding rocket, he gave us his support. We had found a friendly atmosphere within Caltech.

Then it came to the point of whether or not our first program would continue. I decided I would like to do my doctoral thesis on rocket propulsion and rocket flight. I first went to Clark Millikan, and he said, Well, you know, this is a very good time to go into the aircraft industry, and I think you would do better to do that than to go on to a doctorate. That was a rather serious blow for me. So I went to Kármán and he said, It'll be all right, you can do it.

During this first period we had to scrape together whatever money we could from our own pockets to buy equipment. I gave a seminar in April 1937 on the work we had done and that brought us our first financial support. After the seminar, Weld Arnold, who was working as an assistant in the Astrophysics Laboratory at Caltech, came to me and said, could he join our group and could we use $1000? I said certainly, you can be our "official" photographer.

He went away and not too long after came to my office with a newspaper bundle. He said, here's the first $500. When I opened the bundle I found it full of $1 and $5 bills. I never found out how he got them. I took the bundle to Clark Millikan and I said, Clark, how do we open a fund for the rocket research project at Caltech? And he said, What?! I said, Well, I've got $500 and $500 more to come. He was really flabbergasted! What's interesting is that when I checked with the Caltech comptroller in 1946, $300 of that fund was still unspent; it later disappeared into the Caltech general account.

What is called the original Caltech rocket research group was now formed and consisted of Parsons, Forman, Smith, Tsien, Arnold, and me. We were permitted to work through the hospitality and good sense of Kármán. Sad to say, of those six only three are still alive: Amo is here, Tsien is in Beijing, and I am here.

In December 1938 I was asked to give a luncheon talk to the Sigma Xi Society at the Athenaeum. I called it Facts and Fancies of Rockets, and I closed by saying, "The time has come to throw away the pencils and grab a monkey wrench," because we thought that there was enough theoretical understanding to get something done practically if means were available. At the end of the lunch, Robert A. Millikan, Kármán, and Max Mason came to me and said, We'd like you to go to Washington to give expert evidence on rocket propulsion to the National Academy of Sciences committee for Air Corps research. This was a sign that support for our research might be found. Some months before that Smith had gone to work at Douglas Aircraft, Tsien had shifted to other work at GALCIT, Parsons and Forman were back in the desert making explosives, Arnold had vanished we knew not where, and I was left alone—still writing my thesis. I did not finish that thesis until 1940, by which time Kármán was director and I was chief engineer of the Air Corps Jet Propulsion Research Project, GALCIT Project No. 1.

There had been an inkling that interest was rising in rocket propulsion when Kármán was on the East Coast several months before, visiting officers of the Army Air Corps. Also, I was called to San Diego by Reuben Fleet, then president of Consolidated Aircraft Company. He said that he had heard that there were some fellows at Caltech doing something with rocket engines, and he wondered if they could not be used to assist the take-off of flying boats. I had a very nice lunch with him, during which he said, You know, I think that a rocket engine will work much better under water than it does in the air. And I said, well, that doesn't sound right because it just doesn't fit with what we understand about rocket propulsion—a rocket engine works best in empty space, in air a little bit worse, and in water worse still.

Well, he said, come on and we'll make an experiment. We went out into a little garden next to his office where there was a fish pond. He got a hose with a nozzle and said, you hold the hose, and I'll turn on the water so you can feel the thrust of the water jet. Then put the nozzle in the fish pond and see if you don't feel more thrust. I said, Honestly, I can't tell the difference; but he was convinced that there was more thrust; of course, there actually was less.

At this time I was taking Fritz Zwicky's course in analytical mechanics. In my thesis I had run into something I did not understand, so after one of the classes I went up to the famous astronomer and I said, I wonder if you could help me. And he said, What are you doing? Some of you knew him. He was rather brusque at times. When I explained to him what I was doing he said, Well, you know, you're a bloody fool. Rockets have to push on air to operate to get any thrust. I knew there was no point arguing with him. So I just said thank you very much, Professor Zwicky, and left. Two years later he was a consultant for our Project in the Arroyo and shortly thereafter was made director of rocket research at the Aerojet Engineering Corporation—a JPL offspring. He had changed his mind!

As a result of my presentation to the National Academy committee, Kármán, Millikan, and Caltech were offered a choice of five problems on which the Air Corps wanted research done. One of them was aircraft auxiliary rocket power; another was the de-icing of windshields. Jerome Hunsaker of MIT said to Kármán, You can have the Buck Rogers job; I'll take the de-icing of windshields. But for this accident of history, JPL might well be at MIT.

We got a $1000 contract to make a study of rocket literature and to prepare a research proposal for $10 000. The National Academy of Sciences decided it was time to hand the work over to the Army Air Corps. We called the new project the Army Air Corps Jet Propulsion Research Project. It was at this time we had to drop the use of the word *rocket.* Rocket was then a bad word among "serious" people, a science-fiction word, so the Air Corps had adopted the term of "jet propulsion" instead of "rocket propulsion." That is why we now have the names Jet Propulsion Laboratory and Aerojet General Corporation.

Our original rocket research group at GALCIT had been also carrying out theoretical studies of the flight performance of sounding rockets. The group chose as its first goal the design and construction of a sounding rocket—in the backs of our minds we dreamt of voyaging to the moon. When I was a boy of twelve, I read Jules Verne's 1865 novel *From the Earth to the Moon* and never forgot it. But fate determined that we would first go into the development of engines for the propulsion of aircraft, which turned out not to be a bad step, because once one had practical engines, sounding rockets and space vehicles could be built.

The sounding rocket that we had hoped to build did not materialize until 1945 at the Jet Propulsion Laboratory. I realized when returning from a mission to England during World War II that we had now in our hands at JPL all that was necessary for making a successful sounding rocket. I stopped over in Washington, went to the Army's Ordnance Department, and asked the authorities there (things were much easier than I gather they are now), Could we make a sounding rocket? Oh, they said, that's an interesting idea. After checking with the Signal Corps, they said to go ahead—but it must reach an altitude of at least 100 000 ft.

When I returned to JPL I met with Homer Joe Stewart and other research engineers to plan the rocket. Within nine months we tested the WAC CORPORAL in New Mexico, and it reached about 240 000 ft. Nine years had elapsed since our group pinned on a wall at GALCIT a chart outlining the components needed for a sounding rocket; nine years seemed much longer then.

F. J. Malina with WAC CORPORAL at White Sands, New Mexico, November 1945. (Courtesy Caltech Engineering & Science Magazine.)

It is interesting to point out some of the things that were done by the brain-children of GALCIT, primarily in rocket propulsion. The first really important thing was that Kármán and I proved, I am sure for the first time, that it was possible to design a long-duration solid-propellant rocket engine. Up to that time there had been powder rocket engines whose burning time certainly was not more than about 3 sec; they were invented about 800 years ago. There was a firm, sort of intuitive conviction that it was impossible to make a solid-propellant rocket engine that would provide thrust for 10, 20, or 30 sec.

Kármán and I proved theoretically that it could be done, and as a result, the so-called JATO rocket engine came into being with the help of no one else but the explosives chemist Jack Parsons. I am sure he is responsible for the development of what are now called composite propellants, because the first one that we developed was a mixture of potassium perchlorate and asphalt, and I think it unlikely that anyone would have thought of using asphalt as a fuel except someone like Parsons.

We have with us General Homer Boushey who made the first JATO takeoffs, with an Ercoupe aircraft at March Field, California, in 1941. He is

Capt. H. A. Boushey piloting 'Ercoupe' with jet-assisted take-off, August 1941. In white shirts are F. J. Malina (right) and Homer Joe Stewart. (Courtesy Caltech Engineering & Science Magazine.)

alive and here, and I must say I admired his courage. Our black-powder JATOs delivered 25 lb of thrust for about 8 sec. Fortunately, we transported them "fresh" each morning from the laboratory in the Arroyo to March Field; shortly after the flight test we found they blew up after being stored for a few days. We first made some static tests on the ground. The second time Boushey threw the switch one of the JATOs blew. We were really clever. We had mounted them on rails so that if one blew, its nozzle would soar backward and its chamber forward without damaging the aircraft. The system worked but we had not taken into account the fact that the nozzle would ricochet from the ground in a static test. When Boushey climbed out of the cockpit after the explosion, he confronted a hole in the tail end of the Ercoupe fuselage.

But he did not get discouraged. We repaired the damage and sent him on a test in level flight with one JATO under each wing. We saw a smoke trail in the distance and heard an explosion—another JATO had failed. After a dreadful period of waiting, we saw the Ercoupe returning to March Field undamaged. About 140 JATOs were used in the remainder of the flight tests and none failed. An improved, storable composite propellant JATO was then developed and its descendents boost and propel the large missiles and space vehicles of today.

Parsons and I obtained a patent on spontaneously ignitable, storable liquid propellants; the first combination consisted of red fuming nitric acid and aniline. A variant of this combination is now used in the Titan missile and was used in the Apollo program that took a man to the moon. In April 1942 we carried out flight tests of liquid propellant JATOs on an A-20A military airplane piloted by Colonel Paul Dane, who also is here with us. He had a less trying experience than Boushey, because we had learned a lot about rocket engines by then.

From the first elementary experiments that we made at GALCIT in the 1930s, with time and circumstance, and good luck and good helpers, the Jet Propulsion Laboratory came into being. When I left JPL at the end of 1946 we had a staff of about 350 people. I understand there are now about ten times as many people there.

So I say, hail to those of GALCIT who have gone to the other side of space and good luck to those of us who are still alive.

EARLY SUPERSONICS AND BEYOND

Allen E. Puckett

As you rise through various levels of management, the main requirement seems to be that you learn less and less about more and more, and if you understand monotonic functions, you see the obvious end result: eventually, you know almost nothing about practically everything.

Those days of the 1940s were incredibly exciting. I remember coming here in 1941; I had been offered a job as assistant to von Kármán, and that sounded pretty special. It turned out that he had a lot of assistants, so I really was not all that special. It was at the magnificent rate of pay of $120 a month. The hours were not specified; it might have been two-thirds time, something like that, but apart from whatever I did in the lab, I was supposed to do some studying. I was expected to help complete a small supersonic wind tunnel and then to operate it, experiment with it, and so forth. And I had the great pleasure today, as I wandered around through the basement of Guggenheim, of seeing that same wind tunnel, in almost exactly the same form, the same horrible-looking hardware, and primitive instrumentation. I am delighted that it survived for almost forty years.

The story of that little wind tunnel is rather interesting. I found out that it was being constructed on a government contract, and in those days I knew nothing about government contracts. Times have changed. The contract, as I recall, was for $10 000. Bill Zisch was manager of GALCIT procurement and accounts in the Guggenheim Lab, and he told me that this was the biggest contract that they had ever had, practically a goldmine, and that I should be very careful with this incredible sum. I found also that the contract was really with the U.S. Army Ordnance Corps, which struck me as strange. This was a supersonic wind tunnel. Presumably, we were looking forward to high-speed flight, things like that.

It may have been Kármán who told me this story, and it is true as far as I know. I hope it is not apocryphal. He had gone to his old friend Hap Arnold, in the Army Air Corps, with a proposal to build this small supersonic wind tunnel, and Hap Arnold allowed as to how it was an interesting project, but ob-

---

Dr. Puckett, who was a graduate student and faculty member at GALCIT during 1941-1949, is now chairman of the board and chief executive officer of Hughes Aircraft Co.

viously the Air Corps had no interest whatever. There was no chance that an airplane would ever fly faster than the speed of sound, and therefore the Air Corps was not in that business.

Kármán, who was very ingenious, had a great thought. There were devices that already went faster than the speed of sound—the artillery projectiles used for many years by Army ordnance. So he went to the Army Ordnance Corps. How he sold them this bill of goods I'll never know, because their art was pretty well developed. They needed a wind tunnel like a hole in the head. But somehow he sold them the idea that it would be a whole new world to learn great things about projectiles and how they fly through the air. So that was the beginning of the little 2.5in. wind tunnel. The very first shapes that we made to test were models of artillery shells and projectiles, not of airplanes or rockets or wings.

I should like to re-create for you the sense of excitement we experienced. Almost everything we did was new. One could not turn on the tunnel in the morning without seeing something no one had ever seen before. I remember the first time we turned it on, I thought I was seeing a tropical rainstorm or maybe a blizzard. It turned out that we had forgotten condensation. What we had really made was just one helluva refrigeration system, and we were condensing water and ice. So, of course, we had to put a dryer in and pretty soon it worked.

I won't try to recount all the fascinating new things we found. I recall one instance when we had an optical *schlieren* system. We could look through the test section of the tunnel and it would show up the shock waves, the various regions of changing pressure, etc. One day we saw the region where there should have been a shock wave, but it was a pretty fat shock wave. I thought we had discovered something obviously very important. Nobody had ever heard about a very thick shock wave.

So we dashed up to tell Kármán about it, and he made one of his standard comments—that nothing looks as much like a new effect as a mistake. I think it was Hans Liepmann who suggested that if instead of taking our pictures with a steady arc lamp we would better use a spark, a very rapid exposure, the flow might look different. We did that and found that sure enough there was such a thing as an oscillating shock wave. It led to all kinds of interesting things later on.

Hans's laboratory was next door to the transonic wind tunnel and he was exploring flow in this regime, which was at least as mysterious as the supersonic. The transonic regime in many ways was the most mysterious, because nobody really had any notion as to what took place when one went from below the speed of sound to above the speed of sound. There were theories of the sound barrier and all the terrible things that might happen. Even if you looked at it from a theoretical point of view, all our equations ended up having some kind of term like $1/(1-M^2)$ if you were subsonic, or if you were supersonic something like $1/(M^2-1)$, and obviously all that told you was that as M got very

The 2.5in. supersonic tunnel, 1942. (From NDRC Report A-38.)

close to unity, things got very exciting. So that did not shed much light on the details.

Furthermore, designers of airplanes flying well below the speed of sound had no interest whatsoever in something called Mach number. Nobody had ever heard of the compressibility effect, until the pilots of a few of the very-high-speed airplanes (one of them was Kelly Johnson's P-38) did foolish things like diving straight down, at which point they discovered that strange things happened. The control systems did not work in the normal fashion, in a few cases wings came off, and so on.

At that point, in 1943 or 1944 as I recall, Kelly or someone at Lockheed called Caltech to say, We think we may be encountering something that has to do with compressibility. Nobody over here knows anything about it. Could you send somebody over here to tell us about it? Hans and I found ourselves giving a series of lectures at Lockheed Aircraft to a class that included Kelly Johnson and a few great old names. We presumed to tell them what happened to airplanes at high speeds without really having the faintest idea ourselves, but perhaps we at least shed a bit of light on the mystery. They were very exciting days. Anything we did was new.

Then, of course, a number of things began to converge on that area. There was a means now of making things go faster than sound. Guided missiles had appeared in Germany as a so-called practical weapon. All of a sudden, there was tremendous interest in vehicles that really would fly faster than sound. At the same time, the turbojet engine was evolving in Great Britain, Germany, and the United States. So interest began to develop in wings, stabilizers, and bodies that might resemble missiles or aircraft at supersonic speeds.

At the Aberdeen Proving Ground in Maryland around 1944, I had a call from Kármán who was in Washington (he seemed to spend about half his time in those days in Washington). He said, I have something funny I want to talk to you about; can you come down to Washington and meet me at the hotel? So I went down that afternoon and met him. He said, somebody tells me that the Germans are making airplanes with wings that are called "swept back." The reason is that if you visualize the airflow over this wing at very high speed, only the component of the velocity that is perpendicular to the leading edge affects the pressure distribution. The other component is along the parallel to the leading edge and has nothing to do with the distribution. So, he said, you fool the wing. You make the wing think it is really going slower.

Now, he said, this sound pretty crazy but maybe there is something to it; I think you should compute something. I won't bore you with the details, but in his usual fashion, he had seen a way to do a theoretical computation of wings with funny angles and tilted leading edges, etc. For the first time we began to get some insight into the behavior of wings as they go through the speed of sound and into the supersonic region. It was impossible not to discover something new every week or every month.

In the late 1940s the world suddenly exploded with supersonic wind tunnels, with many papers and books on high-speed aerodynamics; and then came the first airplane that really did fly faster than the speed of sound. I think it was the Bell X-1, built like a brick wall, which did in fact prove that man could penetrate the sound barrier and not suddenly revert to childhood (which was one of the theories). I would not give up the excitement of that decade of the 1940s for anything in the world.

In fact, there was reason to think that we had suddenly reached a new plateau of skill in all the arts relating to high-speed flight. A study was made by Kármán and several colleagues, *Toward New Horizons.* It was a projection of all the good things that would result from this fantastic leap in technology that had taken place in the 1940s. It purported to describe the aircraft of the future and all the things that would emerge from what we learned. It was a massive tome and a great projection, but there were a few things that it did not predict—ICBMs, man in space, transistors, integrated circuits and solid-state electronics generally, communication satellites, or electronic computers, all of which were operational within fifteen years—an interesting commentary on our

JPL wind tunnel: A. E. Puckett (right), Dalton Bergan. (Courtesy Caltech Archives.)

ability to extrapolate from our own art at any given moment into what may be a revolution just around the corner.

Let me give you a look at what I think was the next explosion, which you know all about but perhaps have not put in perspective. I am speaking of the explosion that has taken place in the electronics world, or more broadly its effects on data processing, data storage, and communications—what I call the electronic triangle. Each one of them is important in its own right, but together they have literally changed the face of the world.

For example, you can buy a little pocket computer today, a little programmable computer for a couple of hundred dollars, and put it in your pocket. If you do the analysis on the capabilities of that little computer, it can be compared with one of the standard large IBM mainframes of the middle 1950s, which filled a whole room full of vacuum tubes. It had about the same capacity as this little pocket computer, about the same number of available program steps, and the same memory. However, compared with that old IBM machine, this little thing you put in your pocket is about 100 000 times smaller, uses about 100 000 times less power, costs about 10 000 times less, and is about 10 000 times more reliable.

That is what has happened in twenty years. I know of no other technology where the pace and the rate of change has been even close to that. So this is another incredibly exciting world. It is really a combination of that world with the exciting world of the 1940s that has led to some of the things to which we have alluded today. The gigantic steps in space and what we have learned about how to build spacecraft are really a combination of this electronics art with the arts of jet propulsion and high-speed flight. It puts them all together. When I think of some of our first adventures in space at Hughes I don't know whether to laugh or cry at how little we knew then.

One of our first major programs was called Surveyor, managed by the Jet Propulsion Laboratory. Between the crew of engineers at JPL and the crew of engineers at Hughes we probably had four times as many engineers as we really needed. One of the things we learned was that in order to have the necessary reliability in space, the quality of space hardware has to be completely different from that of airborne equipment. The hardware built for airborne use generally costs about $1000 a pound—and, incidently, that was the figure in the middle 1950s. The only difference is that a pound of electronic hardware today does perhaps a 1000 times as much as that pound did in 1950.

Space hardware is about $10 000 a pound, or maybe $20 000, depending on exactly how you figure it. So we have a factor of at least 10, and sometimes as much as a 100, between the cost of space hardware and the cost of airborne hardware. The only difference really is the requirement for reliability—the demands you put on the selection of components, the tracking of components, the testing of subsystems, the testing of the complete system, the attention to design, etc. Reliability is a feature of the product that has a price tag just like anything else.

First of six Intelsat TV-A telecommunications satellites, launched 1975. (Courtesy Hughes Aircraft Co.)

Today in space we have communication satellites that bring us television live from the Middle East or anywhere around the world, and we take it for granted. The communication satellites that are in orbit today just over the Atlantic Ocean alone have a capacity of something like a thousand times the total communication capacity that existed by cable in 1962. That's a factor of 1000 just in our ability to talk across the water on that one route alone.

In data processing, the magic that has totally transformed electronics is a bit of hardware we all see but perhaps without comprehending what it has done to our lives: the silicon chip. We routinely manufacture this chip with thousands of transistors on something the size of your little fingernail, with all of the circuits designed into it, so that circuit design today has to do with the design of that one little chip, not with a wiring diagram or a block diagram.

We are right in the middle of the explosion, and the ways in which it is affecting our lives are so subtle and pervasive that they are really hard to comprehend. Ernie Sechler commented on the use of big computers in structural analysis and the fact that we are doing problems today that could not be done twenty years ago. In fact, design work in many areas is now done entirely by computer. Forget the drawing board. The designer sits at a cathode-ray tube, moves light pencils around, punches keys to call up design modules that already exist in a memory, puts them on the tube, moves them around, reorients them, changes scale, and invents a new module which he then puts back in the computer memory. Meanwhile, all these data are being stored. When he is through with his drawing, every detail is stored in a computer memory that can then be converted to a tape, which will tell a machine somewhere how to make the part. It will also make up the parts list, the master diagram, and add itself to a master drawing list.

The whole process depends entirely on the computer. The computer allows us to do things today that could not be done at all twenty years ago. Ernie worried about the computer replacing the human mind, distracting us from our analytical ability. Well, that is something to worry about, of course. I think if the computer is improperly used it could do precisely that, it could make us very lazy. But there is another way to think about it—not just the computer by itself, but the computer coupled with communication and storage capabilities. It provides a multiplier to the human mind. It provides to the mind what the lever or the wheel were to our muscles 10 000 years ago. We now have something that expands our ability to cope with problems of logical intricacy far beyond the ability of any one mind to grasp.

This is a very exciting time, and the greatest pleasure in my life has been to see those early days of high-speed flight come together with this newer world, to see the action and the change and the confluence of the technologies that have produced such fantastic things. And having belonged to the GALCIT group, with the sense of teamwork, adventure, and excitement, is something that I shall never forget.

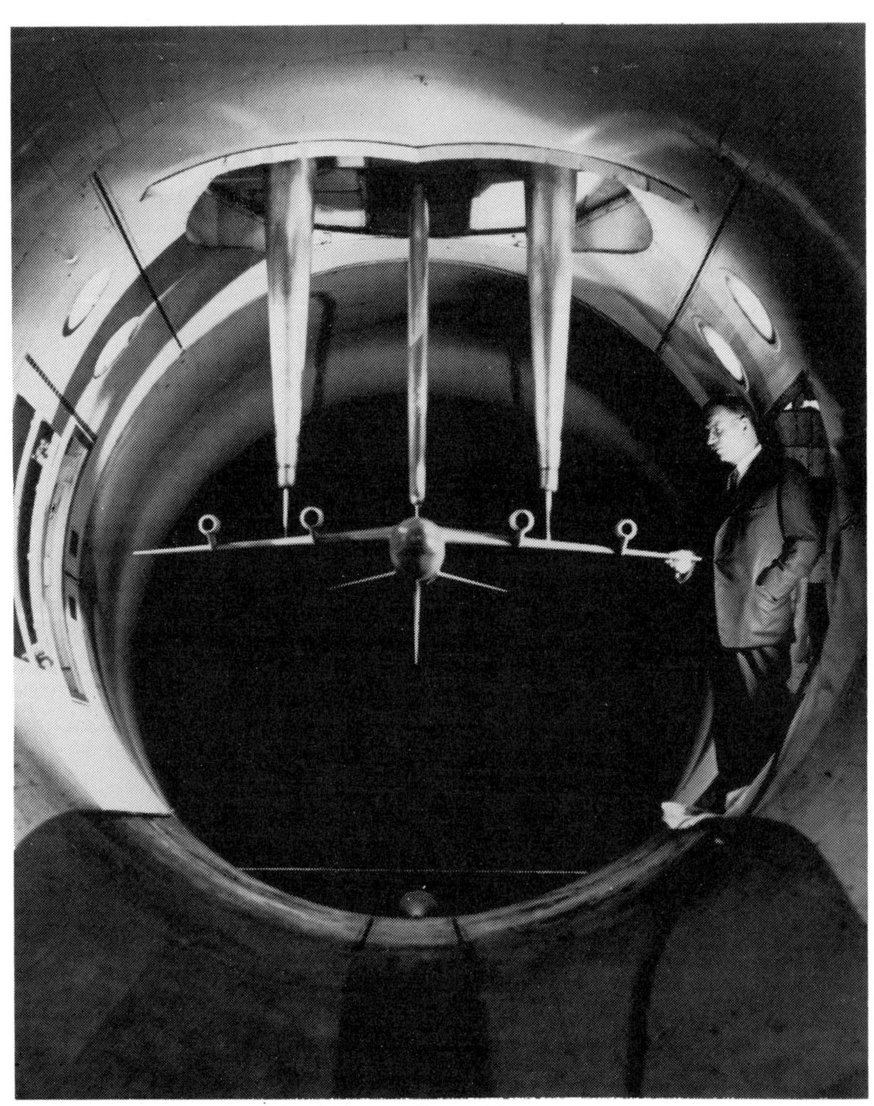

Clark Millikan in 10ft wind tunnel. (Courtesy Caltech Archives.)

Clark B. Millikan in Memoriam

Hans W. Liepmann

No GALCIT festivities would be complete without paying homage to Clark B. Millikan, CBM for short, who spent his entire professional life at GALCIT and was its director from 1949 until his death in 1966.

Even before Kármán came permanently to Caltech, Clark together with Maj Klein and Ernie Sechler did the detailed design and supervised the construction of the 10ft tunnel, for a long time the only large and (for the time) fast wind tunnel on the West Coast. Since its completion in 1929 this tunnel has provided a link with the daily problems of industry that is hard to overestimate. The exchange of ideas between academia and industry was always a strong component of CBM's interest, and both the old 10ft tunnel and later on the cooperative transonic tunnel (CWT) served this purpose.

Besides being an accomplished research worker, Clark was an outstanding classroom teacher and combined in both teaching and research an excellent background in mathematics and physics with a real appreciation for engineering design. He contributed, for example, one of the earliest fundamental papers on airplane performance as well as a classic paper on the fundamentals of turbulent boundary layers. He could not, of course, escape the pressure of administration, of national committee memberships, and of an ever-increasing number of visitors, etc.—the usual penalty inflicted upon a successful and capable academician. With the combination of an iron dedication to what he considered his duties, and very hard work, he managed to keep up with the latest developments in the field. His classes were up to date until his death; his criticism and suggestions on research reports and papers from the GALCIT inmates, poignant and competent.

For me, his outstanding characteristic as director of GALCIT was his generosity in providing means for the GALCIT researchers, furthering them in every way without the slightest trace of jealousy or resentment of being left out of the glamorous part of research. CBM received his share of honors but I doubt that the scientific community as a whole can be aware of his indirect but crucial contributions to GALCIT research, indirect not because of a lack of ability but because of an ingrained dedication to what CBM considered his primary duty.

CLOSING REMARKS

Hans W. Liepmann, A. L. (Maj) Klein, and William R. and Mabel R. Sears

Hans W. Liepmann: I want to mention a few people who wanted to attend the Symposium but could not and sent letters and telegrams. I did send a little note to [U.S. Secretary of Defense and former Caltech president] Harold Brown to say, I know you have something more important to do, but it would be nice if you could be here. I got a very nice note back saying, Yes, that's quite right, and I wish I could be there; thank you very much. Here is a list of people who sent greetings to the whole GALCIT family and friends.

Johann Arbocz, Technische Hogeschoole, Delft

Harold Brown, Secretary of Defense, Washington, D.C.

Rolf D. Buhler, Director of Institute für Raumfahrtantriebe, University of Stuttgart

Lester S. Chambers, Rear Admiral (ret.), St. Petersburg, Fla.

Paul F. Chenea, Vice President, General Motors Corp., Warren, Mich.

Stanley Corrsin, Professor of Fluid Mechanics, Johns Hopkins University, Baltimore

Christopher Dykes, Managing Director, Continental Oil Co., London

Robert D. Fletcher, Chief Scientist, USAF Air Weather Service (ret.), Tubac, Ariz.

Ivan A. Getting, President, Institute of Electrical and Electronics Engineers, New York City

Alan H. Green, Real Estate Investment, Williamstown, Mass.

Robert H. Korkegi, Director of AGARD, Neuilly-sur-Seine, France

Ting Y. Li, Professor of Aeronautics and Astronautics, Ohio State University, Columbus

Hans Mark, Undersecretary of the Air Force, Washington, D.C.

Ruben Mettler, President, TRW, Los Angeles

Franciscus Nieuwstadt, Research Associate, Royal Netherlands Meteorological Institute, De Bilt

Silva Ozires, Chairman and Chief Executive Officer, Embraer Empresa Aeronautics, São José dos Campos, S.P., Brazil

Lillian Price, Gaston, Oregon

Donald L. Putt, Lt. General USAF (ret.), Atherton, Calif.

Eli Reshotko, Professor and Chairman, Department of Mechanical and Aerospace Engineering, Case Western Reserve University, Cleveland, Ohio

Milton Rogers, USAF Office of Scientific Research (ret.), Baltimore

Guyford Stever, President's Science and Technology Advisor, Washington, D.C.

Theodor H. Troller, Ava Ranch, Portola, Ariz.

Elmer F. Ward, Santa Ana, Calif.

Frank Wattendorf, Washington, D.C.

Thomas F. Weldon, Consultant, Boeing Co., Paris, France

Max L. Williams, Professor and Dean of the School of Engineering, University of Pittsburgh

T. A. Wilson, Chairman of the Board, Boeing Co., Seattle

And now I should like to introduce one more person, one of the oldest members of GALCIT, who has been mentioned several times, Maj Klein; and then we shall hear from Bill Sears. As you heard Arthur Raymond remark, everyone at Douglas complained about Maj, but when you asked them whether Maj should leave, they all said, "Oh, God, no!" When I was doing a lot of consulting for Douglas, I noticed that most of the designers had two sets of drawings on their drawing boards. One was actually on the board, and the other they pulled out and put on top of the first when they saw Maj approaching, so that he would feel free to draw all over it and say "No, no—this is all nonsense." I should now like to have Maj say a few words about the old times here.

A. L. (Maj) Klein: I got into the aeronautics department in an unexpected, unplanned, and most improbable manner. I had a laboratory in the physics department where I was doing actually very little at the time, but my Ph.D. was in physics, and I happened to have an income at the time which made it unnecessary for me to get a job, so I was just puttering around in that laboratory. Clark Millikan was across the hall, and he and a classmate of his named Hervey Hicks, who was a mathematician, were trying to design equipment for GALCIT and the Merrill Bright airplane. I went in and looked at things like the landing gear and said, "You can't drill that hole in that direction, the landing gear won't work," and various other simple things, because neither one of them had any machine design experience. Soon they started coming across the hall, and I would go over there and straighten them out, and I would do this and I would do that.

Then, pretty soon it was a case of designing the wind tunnel, and they were not sure how to put the vanes in the corner, so I put the vanes in the corner. Then came summer. At that time three men were working on the Merrill plane; a building was going up with the GALCIT wind tunnel in it and everybody in the aeronautics department officially went on vacation. Then the foreman came over and said, "How do you do this?" and "How do you do that?" and I would tell him, and nobody seemed to object so I went on telling him.

All this time I was still an unpaid research fellow in physics. Afterwards, I guess they decided I had better be put in the aeronautics department because somebody might question what I was telling people to do, and seeing that I was the only one around to tell them, why they had better give me a little authority. So that is how I got into the aeronautics department, two years after I had my Ph.D. in physics; and of course, I was always more interested in machine design and things like that than I was in the more obscure parts of physics, so it was just a natural opportunity for me to get my fingers into a lot of complicated jobs that nobody else seemed to want to do.

I was given the job of designing the rigging for GALCIT, so I designed the rigging for GALCIT. They were happy with three degrees of freedom; but eventually they had to have six, so I designed it for six. They gave me a series of loads, a 1000-lb model and 1500-lb load. The first model that came in weighed 4000 lb and had 4000 lb of load. It was a 40% scale model of the A-26 and had a 400hp electric motor in it; with the crosstunnel model, we could fortunately put on counterbalances, so that we could carry its weight and still measure what happened to it.

Then in Co-op tunnel (I had the job of designing that one too) they told me they were going to have 30 000 lb of lift, so I designed a system for 30 000 lb of lift, which was pretty heavy, of course; but they never got more than half of that, because when they had this great big crosstunnel model at maximum angles of lift at 450 mph, it shocked, and so they had shock phenomena in the wind tunnel. They never could get it up to full speed, and the loads never got within a factor of two of what we had designed for, but they still wanted to measure those little loads of 1% down to practically nothing. To do that we had to put in tilt meters and measure the tilt of the earth due to the earth tides from the tides in Santa Monica. Now that's an honest fact. We finally ended up being able to measure them, but I do not think we ever did it.

That was my experience as a GALCIT machine designer; most of these things seemed to work pretty well, so I went on talking and arguing with people about this and that, and even though their ideas seemed pretty queer at times, usually we would work out some way of solving problems. Kármán put some tough ones up to me, but I managed to get through them somehow. That is about all I can say about my endeavors in the early days of GALCIT. The experience was very rewarding to me personally; I enjoyed doing what I was doing, and I enjoyed the people with whom I was working.

William R. Sears and Mabel R. Sears: I want to begin with the observation that GALCIT during the Kármán years from the late 1920s until 1941—in that really very short time—became an exceedingly prestigious and world-famous institution. It seems profitable to look at such an institution and see what it really was. If you can put your finger on them, what were the qualities, the peculiarities of that institution, that brought it to such eminence in that short time? In that period, GALCIT achieved an international reputation as a research center and as a training ground. The world's most distinguished fluid mechanicists came here on short-term and long-term visits; its publications and communications were many, and almost dominated the literature of the field; its students went out into government, industry, and universities to become outstanding engineers, scholars, and captains of industry. The influence of this small institution was really remarkable; so it is not only pleasant to reminisce about those delightful years, but it is worth while to try to see what caused it.

As you have already heard from Maj, and also from Hans Wolfgang [Liepmann], fluid mechanics did not begin with Kármán's arrival; it was already here. Bateman was actually in Throop College before the institution became the California Institute of Technology, and Albert A. Merrill, whose name has been mentioned a couple of times, a research assistant, had built a wind tunnel and was in charge of courses in aeronautics. Maj has pointed out that this little team, formed and really I think sparked by Merrill, was interested in the "Merrill plane" and in Merrill's wind tunnel. Clark Millikan came from Yale and got interested in aeronautics and, as Maj has told us so clearly, neither Bateman nor Merrill nor Clark Millikan knew how to design anything, so they enlisted Maj to help them with the airplane and with the wind tunnel. So, the "Merrill plane" was built and flown, and there was a wind tunnel; but the important thing is that this team, which I see as Merrill, Bateman, Millikan, and Klein, had been drawn together at a most propitious time, because the Daniel Guggenheim Fund for the Promotion of Aeronautics had been formed and was accepting proposals.

Now, everything was breaking right. Clark Millikan's father was the chairman of Caltech and it probably was not too difficult for Clark to convince his father that they ought to look into this thing. We have heard some of the details; it was done very astutely, and with the cooperation and encouragement of the burgeoning southern California aircraft industry. So Caltech's proposal was accepted. A feature of Caltech's plan was to bring in an outstanding director, and Hans Wolfgang has told us about the remarkable letter that Robert Millikan wrote to Harry Guggenheim, mentioning the three youngsters he thought might be prospects for this position—Kármán, Ludwig Prandtl, and Geoffrey I. Taylor (there is some talent in there some place!); and then saying clearly and distinctly that, after thinking it all over, on the basis of age and personality and all considered, Caltech thought that Kármán would be the one to choose. It was Paul Epstein who brought Kármán's talents to the attention of the chairman. So the job was offered to Kármán and he accepted.

Kármán first visited Caltech in 1926, and the first thing he undertook was to sit down with Maj and Clark and talk about the wind tunnel that was planned for this Guggenheim Laboratory. They had already made a plan for a large, open-return wind tunnel, but Kármán convinced them that the tunnel that we see there now, the 10ft closed-return Göttingen type, as it was called in those days, in contrast to the flowthrough Eiffel type, would be more suitable and more efficient in the use of space. So the new design with the circular, 10ft closed-working section formed the core around which the Guggenheim building was built, with offices and laboratories. Of course, at this point you can twit Maj a bit about the rest of what they forgot in the building; anybody who sees those corridors and that little steel staircase—only one up the front there—can see clearly that they did not really expect to have any students there. There was only one small classroom, Room 303, and we called it the Seminar Room; it held about thirty-five people.

From the start, if you read the proposal they made to the Guggenheims and you look at the building they built to house the laboratory, it was clearly conceived as a small institution. There were about five faculty members; you can count the five and when you get to me, that is about the sixth, and I don't know if there is a message in there or not. We made do with a minimum of office space, and the outfit began its active existence as GALCIT in the fall of 1928. In October 1929 Kármán agreed to become the director. When he had come earlier he had not yet clearly made up his mind and came as a consultant on the wind tunnel and the building; after having consulted with them, having been here and seen Pasadena, he moved to Pasadena with his mother and sister in December 1929. He continued to hold the position of director in Aachen, dividing his time equally between the two institutions until 1933, when he resigned from Aachen and came full time to Pasadena.

It is clear that Kármán brought to this institution two characteristics that established the spirit and the atmosphere which pervade it even today. The first was his emphasis on original research and publication, which I think we can recognize as a natural carryover from his Göttingen-Prandtl background; but in the 1920s and early 1930s it was not exactly the common characteristic of American engineering colleges. The second quality was his informality, which seems to have been an inherent trait of the director himself and was hardly typical of the Central European background from which he came; it was certainly well received and encouraged by Caltech's informal atmosphere and the informal proclivities of Clark Millikan and Klein and Sechler.

The weekly research conference was initiated by Kármán. Now, "research conference" means various things in various institutions. Ours was held every Tuesday, and by God it was held every Tuesday! When Kármán was away, Clark, Maj, or Ernie chaired the meeting but it was never, ever skipped. When I first came here, the research conference was a weekly meeting in which people gave a brief statement of their progress or lack thereof, and why, and why not. It was not a conference in which people tried to solve problems. It was a matter of finding out whether people were getting on with their research, how they were getting along, and whether anyone was ready to publish; it was a matter of who has got the equipment that somebody needs, of planning shop time, locating and assigning instruments, and so forth. Kármán had to get a one-sentence statement of this progress, or lack of same, for each guy at the end of the meeting. For one or two weeks it was OK to say, I haven't anything new—same report as last time—but if too many weeks went by, that became more serious. The visiting scholars, the professors, the postdoctoral personnel, and all the graduate students engaged in research reported regularly and made this kind of statement. And it was fun!

The spirit of the institution was really there in that research conference, because everybody was excited and enjoying everybody else's research; and clearly, what was necessary was to let your hair down and not keep anything close to your chest. You had to tell what your troubles were and what you thought was good, expose your ignorance and your errors before your teachers (which is not too easy, as you know) and before your colleagues. The reason it worked was Kármán's objectivity and friendliness. I can remember Kármán grinning from ear to ear when he discovered that somebody's project could be expressed in confluent hypergeometric functions à la Whittaker and Watson. He thought it was hilarious. This informal atmosphere of real joy when things were going right was infectious; it was characteristic of the institution.

Hans W. Liepmann: In closing, I should like to add a few remarks about GALCIT's future.

I am always amused when discussions of space pioneering work and of the men who predicted space flight fail to mention the transistor. Without the transistor and its innumerable applications, real space exploration would have been impossible. The ultimate importance of a research and technology result is quite often hidden and very frequently most surprising. The planning of future research is hardly a boundary-value problem with prescribed initial and final states, but most often an initial-value problem with a given initial direction. Overall aims and goals exist, of course, but they are usually of sweeping generality, encompass a variety of possible technologies and approaches, and do not lend themselves to planning the tactics of the next few years. The limited resources for an increasing population is clearly one such large, even overwhelming, issue.

For the narrower task of the future of GALCIT, I have little doubt that the combination of electronic and mechanical systems will become more and more important. Active controls to make an unstable mechanical system stable as part of a combined system are already in use. Some aspects of active control of fluid flow by means of electronic feedback systems are in the works today. The so-called "smart" wind tunnel which corrects itself is an example. The most important problem here seems to me turbulence control. There are problems such as combustion and chemical lasing where increased mixing and hence turbulence are beneficial; there are other problems such as drag reduction and noise reduction where the decrease or elimination of turbulence is sought. Problems of this kind occur in an enormously large number of vortex applications, from paper making to jet aircraft. Here, then, is one prospect for a future contribution from GALCIT.

Nonlinear waves—whether surface waves on water, shock waves, or light waves in matter—demonstrate a wealth of new phenomena, new technology, and fruitful new ideas to other scientific fields. The solutions of water waves have migrated to particle physics, shock waves have successfully invaded chemistry and materials science, and nonlinear optical phenomena produce almost daily new effects and find new applications. Here is another broad field in which GALCIT participated for many years and which is still expanding at a surprisingly large rate.

Vortex rings once studied as models for atomic structure have returned as important building blocks of complex turbulent flows, as excitation models for superfluids, and as models for some plasma flows. The interaction of vorticity and of vortex-like structures is alive and likely to remain so for some time to come. It is certainly of immediate importance for maneuvering aircraft.

The expected revolution in structural materials is still in its infancy. Composites are certainly here to stay, but here, as in earlier fluid mechanics, I look forward to the closing of the gap between the macroscopic and microscopic approach, between solid-state physics, structural mechanics, and engineering. I believe we are moving toward a time when materials with specified mechanical properties can be made to order; when strength, rigidity, fatigue life and crack propagation characteristics can be designed into a material.

I expect GALCIT to move in these directions. Not in a straight and easy line—that only happens with trivial problems—but with a determination to understand the obstacles in the way and to find ways to overcome them. I have not the slightest doubt that we shall find the future at least as exciting as the past.

Name Index